SpringerBriefs in Mathematics

SpringerBriefs in Mathematics showcases expositions in all areas of mathematics and applied mathematics. Manuscripts presenting new results or a single new result in a classical field, new field, or an emerging topic, applications, or bridges between new results and already published works, are encouraged. The series is intended for mathematicians and applied mathematicians.

More information about this series at http://www.springer.com/series/10030

Steven T. Dougherty

Algebraic Coding Theory Over Finite Commutative Rings

 Springer

Steven T. Dougherty
Department of Mathematics
University of Scranton
Scranton, PA
USA

ISSN 2191-8198   ISSN 2191-8201   (electronic)
SpringerBriefs in Mathematics
ISBN 978-3-319-59805-5   ISBN 978-3-319-59806-2   (eBook)
DOI 10.1007/978-3-319-59806-2

Library of Congress Control Number: 2017943819

Mathematics Subject Classification (2010): 11T71, 94B05

Printed on acid-free paper

This Springer imprint is published by Springer Nature
The registered company is Springer International Publishing AG
The registered company address is: Gewerbestrasse 11, 6330 Cham, Switzerland

For Kelly, Steve and Checco.

Acknowledgements

The author is grateful to Jessica Hollister, Meg Hudock, Josep Rifà and Mercè Villanueva for their helpful comments on an early version of the text.

Acknowledgements

Contents

Chapter 1
Introduction

In this chapter, we give a brief introduction to the history of algebraic coding theory and give the basic definitions and notations necessary to begin a study of the subject.

1.1 History

Coding theory arose in the twentieth century as a problem in engineering concerning the efficient transmission of information. Its study originated in the landmark papers by Shannon [14] and Hamming [9]. Specifically, the theory was developed so that electronic information could be transmitted and stored without error. Electronic information can generally be thought of as a series of ones and zeros. Therefore, coding theory, from this perspective, was largely done using the binary field as the alphabet. However, the alphabet was quickly generalized to finite fields, at least for mathematicians, since many of the proofs and techniques were identical to the binary case viewed as the field with two elements. This type of coding theory remains a vital part of electrical engineering in terms of ensuring effective communication in telephones, computers, television, and the internet.

From the very beginning of its study, mathematicians viewed coding theory not only as an application to electrical engineering and computer science, but also as a part of pure mathematics. They were interested not only in the fundamental questions of coding theory, but also into its connections with other areas of discrete mathematics. Early results connected codes to designs, lattices, and combinatorics. These connections were generally made with codes where the alphabet was a finite field. Moreover, much of the early work from mathematicians in coding theory came by applying previously known results from linear algebra, finite geometry, algebra, and combinatorics to the study of codes. Since the inception of coding theory in 1948, there has been a very fruitful interchange from pure mathematics to the application of

© The Author(s) 2017
S.T. Dougherty, *Algebraic Coding Theory Over Finite Commutative Rings*,
SpringerBriefs in Mathematics, DOI 10.1007/978-3-319-59806-2_1

codes. As often happens in applied mathematics, interesting mathematical questions arose in the application which sparked mathematicians' interest in the subject.

During the first forty years of coding theory, the alphabet in question was usually a finite field. There were a few papers written where the alphabet was a ring, for example Blake's early papers [3, 4]. It wasn't until the 1990s when coding theorists began to study codes over finite rings in earnest. This study began with the understanding that certain non-linear binary codes, which had some of the properties of linear codes were, in fact, the images of codes over $\mathbb{Z}_4$ under a non-linear map. This breakthrough came in [10, 11], however, Delsarte's work in [6], years before, might have led the coding theory community to these results earlier. This prompted an intense study of codes over $\mathbb{Z}_4$ which rapidly moved into the study of codes where the alphabet was either one of the three other commutative rings of order 4 or the ring $\mathbb{Z}_k$. Families of rings presented themselves for study and a large literature emerged studying codes over rings. The families of rings were usually chosen for some specific application. For example, codes over the family of rings $\mathbb{Z}_{2k}$ were studied because of an interesting connection to unimodular real lattices, see [2]. It was a natural generalization from this family to the family of chain rings. Later, with the understanding that all finite commutative rings were the direct product of local rings via the Chinese Remainder Theorem, codes over local rings were studied.

In [16], J. Wood showed that both MacWilliams theorems held for the class of Frobenius rings. This showed that coding theory can be studied over this fairly large family of rings without losing the fundamental foundations of coding theory. Generally, when studying codes over rings, a blanket assumption is made that all rings serving as alphabets for codes are finite Frobenius rings. An extensive and expanding literature now exists on codes over various families of rings.

In this book, we shall not describe coding theory as a branch of engineering, nor shall we motivate its study in terms of communication applications. Rather, we view coding theory as a branch of pure mathematics serving as its own motivation for study. We shall refer to this branch of pure mathematics as algebraic coding theory (which it has often already been called) to distinguish it from coding theory as an application in electrical engineering. Algebraic coding theory sits partly in algebra, number theory, finite geometry, and combinatorics. As such, it has interesting connections to a wide variety of topics in all these branches.

The interested reader can consult MacWilliams and Sloane's seminal text "The Theory of Error-Correcting Codes" [13] for an early description of classical coding theory. For an updated description, see Huffman and Pless's "Fundamentals of Error-correcting Codes" [12]. For a description of the connection between designs and codes see Assmus and Key's "Designs and their Codes" [1]. In all three of these classic texts, codes are generally defined over finite fields.

In this text, we shall be concerned with codes over finite commutative Frobenius rings as was first established in [16]. It will become apparent why we need to restrict to Frobenius rings when we discuss the MacWilliams relations in Chap. 3. We shall give foundational results for algebraic coding theory and develop the structures to view it as an interesting branch of pure mathematics. It is also possible to study codes over non-commutative rings, but much of the theory is different and as yet has not been as widely studied. Codes over infinite rings have also been studied, but generally in close association with codes over related finite rings. For example, codes over the infinite ring of p-adic integers were studied but largely in relation to codes over the finite ring $\mathbb{Z}_{p^e}$. For examples of the study of codes over these infinite rings, see [5, 7].

While we are making the case for algebraic coding theory to be viewed as a branch of pure mathematics, we strongly believe that the bridge to applications must also be open. In fact, we look to applications as a rich source of interesting questions and ideas.

1.2 Definitions and Notations

We shall now give the necessary definitions to begin our study of algebraic coding theory. While we usually choose an alphabet for codes with an algebraic structure, for example a group, ring, or field, some very interesting results can be obtained while simply taking any set as an alphabet. This is how we shall start by taking the most general definition of a code.

Definition 1.1 Let A be any finite set. A code C over the alphabet A of length n is a subset of A^n.

In terms of classical coding theory, the elements are called codewords and the underlying set A is called an alphabet since a code was generally used to transmit information. We retain this nomenclature and we say that if A is an alphabet then C is a code over A. Using this definition one can even think of the English language as a code over the standard English alphabet (allowing for padding of words with blanks to make them all have the same length). The aims of classical coding theory are not that much different than a spell checker to the English language as a code. As we all know, it is quite difficult to have an effective spell checker because the difference between two words in this language might simply be one letter. This idea of distance is applied to classical codes in the same way.

The principal distance used in coding theory is known as the Hamming distance.

Definition 1.2 Let $\mathbf{v}, \mathbf{w} \in A^n$ where A is any set. Then

$$d_H(\mathbf{v}, \mathbf{w}) = |\{i \mid v_i \neq w_i\}|. \tag{1.1}$$

The minimum Hamming distance of a code C is $d(C) = \min\{d_H(\mathbf{v}, \mathbf{w}) \mid \mathbf{v}, \mathbf{w} \in C, \mathbf{v} \neq \mathbf{w}\}$.

We often remove the C from the notation and simply refer to the minimum Hamming distance as d. It is an easy exercise to see that the Hamming distance is a metric on the space A^n.

This definition leads naturally to the following definition. Denote the all-zero vector by $\mathbf{0}$.

Definition 1.3 Let $\mathbf{v}$ be a codeword in A^n where A is any alphabet. Then the Hamming weight of $\mathbf{v}$ is

$$wt_H(\mathbf{v}) = |\{i \mid v_i \neq 0\}|. \tag{1.2}$$

The minimum Hamming weight of a code C is $\min\{wt_H(\mathbf{v}) \mid \mathbf{v} \in C, \mathbf{v} \neq \mathbf{0}\}$.

The fundamental question of classical coding theory is the following.

Question 1.1 Given an alphabet, a length n, and a size of the code M, what is the largest d for which a code exists with these parameters?

Of course, the question can be rephrased in numerous ways by switching which parameter you want to optimize. While much is known about this fundamental question, especially for specific values, the general question remains open. Neither an exhaustive theorem nor an effective algorithm has yet been found to answer the question for an arbitrary set of parameters.

Example 1.1 A Hadamard matrix of order n is a matrix with elements from the set $\{1, -1\}$ such that $HH^T = nI_n$. It follows from the definition that any two distinct rows split evenly between coordinates where they agree and coordinates where they disagree. Let C be the code consisting of the rows of a Hadamard matrix. The code has length n, cardinality n, and minimum distance $\frac{n}{2}$. For example, the Hadamard matrix of order 4 is:

$$\begin{pmatrix} 1 & 1 & 1 & 1 \\ 1 & -1 & 1 & -1 \\ 1 & 1 & -1 & -1 \\ 1 & -1 & -1 & 1 \end{pmatrix}. \tag{1.3}$$

This gives a code with 4 elements of length 4 and the distance between any two elements equal to 2. Except for $n = 1, 2$, the order of a Hadamard matrix must be a multiple of 4. It remains an open question whether there exists a Hadamard matrix for all $n \equiv 0 \pmod 4$.

Definition 1.4 A code of length n, size M, and minimum Hamming distance d is said to be optimal if it is has the largest d of any other codes with length n and size M.

In general, the main question of coding theory is finding optimal codes for a given set of parameters. In terms of actual coding theory applied to electronic communication, we are, in general, not simply looking for an optimal code, but an optimal code which has an efficient decoding algorithm.

This definition of a code requires no algebraic structure. Rather it is simply a combinatorial structure with very few constraints. However, it is enough to get to one of the most important bounds for a code, which was first proven by Singleton in [15].

Theorem 1.1 (Singleton Bound) *Let C be a code of length n over an alphabet of size q with minimum Hamming distance d. Then $log_q(|C|) \leq n - d + 1$.*

Proof Consider the first $n - (d-1)$ coordinates. These must all be distinct, otherwise the distance between two vectors would be less than d. Hence $|C| \leq q^{n-(d-1)}$. This gives that $log_q(|C|) \leq n - d + 1$. $\square$

The most natural application of this result is when $|C| = q^k$, when we have $k \leq n - d + 1$.

Definition 1.5 A code of length n over an alphabet of size q with $|C| = q^k$ and minimum Hamming distance d satisfying $k = n - d + 1$ is said to be a Maximal Distance Separable (MDS) code.

Example 1.2 A Latin square is an n by n matrix with entries from a set of cardinality n such that each row and each column contains each element from the set exactly once. Two Latin squares L, M are orthogonal if the set $\{(L_{ij}, M_{ij})\}$ contains each ordered pair exactly once. Let L and M be a pair of orthogonal Latin squares of order q where the symbols used are $\{1, 2, \ldots, q\}$. Let $C = \{(i, j, L_{ij}, M_{ij}) \mid 1 \leq i, j \leq q\}$. Then $|C| = q^2$, $n = 4$, and the minimum distance is 3. Then $2 = 4 - 3 + 1$ and the code is an MDS code of length 4. It is well known that an orthogonal pair of Latin squares of order q exist for all integers $q > 2, q \neq 6$.

Note that the code in this example is not described algebraically and in fact may have no algebraic structure. Finding MDS codes is largely a combinatorial problem. However, as we shall see later in Example 1.6, one can construct some MDS codes algebraically. Moreover, MDS codes have natural connections to finite geometry and to many open questions in mathematics. We can now state one of the most interesting open questions of the text.

Question 1.2 For which q, n and k do there exist MDS codes?

It is well known that the existence of a finite projective plane is equivalent to the existence of a complete set of MOLS (Mutually Orthogonal Latin Squares). Moreover, a set of MOLS gives an MDS code as in Example 1.2. Therefore, if one were to solve Question 1.2, one would also determine when finite projective planes exists and how many MOLS or order n there are for each n. Since these questions, in themselves, have been open for centuries it becomes apparent how difficult a problem this is.

Another very important combinatorial bound is the following sphere packing bound. We denote the binomial number of choosing s objects from n by $C(n, s)$.

Theorem 1.2 (Sphere Packing Bound) *Let C be a code of length n over an alphabet of size q with minimum Hamming distance $2t + 1$. Then*

$$|C|\left(\sum_{s=0}^{t} C(n, s)(q - 1)^s\right) \leq q^n. \tag{1.4}$$

Proof Let **v** be a vector in A^n where A is an alphabet of size q. There are $C(n, s)(q - 1)^s$ vectors in A^n that have Hamming distance s from **v**. This is because there are $C(n, s)$ ways of choosing the s coordinates to change and $q - 1$ choices for each coordinate. A sphere of radius t consists of all vectors that are distance less than or equal to t from the vector **v**. It follows that there are $\sum_{s=0}^{t} C(n, s)(q - 1)^s$ vectors in a sphere of radius t around **v**.

Given that the minimum distance is $2t + 1$, we have that all of the spheres are distinct. Hence the number of vectors in all of the spheres must be less than or equal to the number of vectors in the ambient space which is q^n. $\square$

Definition 1.6 If C is a code of length n over an alphabet of size q with minimum Hamming weight $2t + 1$ and $|C|(\sum_{s=0}^{t} C(n, s)(q - 1)^s) = q^n$, then the code is said to be perfect.

For a perfect code, the spheres of radius t, where the minimum distance is $2t + 1$, contain all of the vectors of the ambient space. In fact, each codeword in the ambient space is in exactly one such sphere. This is the reason that the code is called perfect.

Example 1.3 Consider the binary code C of length 7 which consists of the all zero codeword, the all one codeword, the characteristic function vectors of the lines of the projective plane of order 2, and the characteristic function vectors of the hyperovals (compliments of lines) of the projective plane of order 2. The code C has length 7, $|C| = 16$, and minimum distance 3, giving $t = 1$. Then $|C|(\sum_{s=0}^{t} C(n, s)(q-1)^s) = 16(1 + 7) = 2^7$ and hence the code is perfect.

Notice again that this code is described combinatorially, but this also has an algebraic description and is a member of a family of perfect codes as seen in Example 1.7.

The proof of the Sphere Packing Theorem shows how codes are used to correct errors. Specifically, a received vector can be any vector in the space A^n. The codewords are the centers of the spheres. Then the vector is decoded to the center of the sphere it lies in (provided that it lies in a sphere). This is precisely why it is desired that the spheres be non-intersecting and cover as much of the ambient space as possible. While this description is completely combinatorial, in practice the algorithms for decoding are highly algebraic.

If the minimum Hamming distance of a perfect code is $2t + 1$, then a vector with t or fewer errors will remain in the non-overlapping spheres of radius t. Moreover, any vector in such a code must be within distance t from a codeword. This is why often in the literature a perfect code with minimum distance $2t + 1$ is called a perfect t-error correcting code.

We shall now begin to study codes from an algebraic standpoint. As such, we are looking for alphabets with an algebraic structure. For us, this structure will almost always be a finite commutative ring. For the remainder of the text, by ring we shall always mean a ring with unity. We can now make the definition of a linear code over a ring.

Definition 1.7 Let R be a finite ring. A linear code C over the alphabet R of length n is a submodule of R^n.

Notice that we are not saying that any module will be a code, but rather only those modules which are submodules of R^n. If R is a field, then the linear codes are vector spaces and we have the full force of linear algebra at our disposal. Given the main problem of coding theory, we are usually searching for optimal codes or codes with some particular characteristic. As such, we are generally concerned with codes up to equivalence given by the following definition.

Definition 1.8 Two codes C and C' in A^n, where A is any set, are said to be permutation equivalent if C' can be obtained from C by permutation of the coordinates. Two codes C and C' in R^n, where R is a finite ring, are said to be equivalent if C' can be obtained from C by a combination of a permutation of the coordinates and multiplication of a coordinate by a unit in the underlying alphabet.

The second definition requires that the alphabet be a ring so that the notion of unit is defined. It is immediate that if C is an optimal code, then every code equivalent to it is also optimal. We shall now show some of the strengths of using linear codes.

Theorem 1.3 *If C is a linear code over a ring R, then the minimum Hamming distance and the minimum Hamming weight are equal.*

Proof Let C be a linear code with minimum Hamming distance d_1 and minimum Hamming weight d_2. Since the code is linear, we have that $\mathbf{0} \in C$. Let $\mathbf{v}$ be a codeword with minimum Hamming weight d_2. Then $d_H(\mathbf{v}, \mathbf{0}) = d_2$. Hence there are vectors that are distance d_2 apart so $d_1 \leq d_2$.

Now assume that $\mathbf{v}$ and $\mathbf{w}$ are vectors such that $d_H(\mathbf{v}, \mathbf{w}) = d_1$. Then $wt_H(\mathbf{v}-\mathbf{w}) = d_1$ and we know $\mathbf{v} - \mathbf{w} \in C$ since C is linear. This gives a vector with Hamming weight d_1 so $d_2 \leq d_1$. Hence, we have that $d_1 = d_2$. $\square$

Since the minimum distance and minimum weight are the same for linear codes, they are often both denoted by d. Notice that, as we have just done, the term Hamming from minimum Hamming distance and minimum Hamming weight is often dropped. It is usually only used when there is ambiguity about which weight is being used.

For codes over finite fields, we often denote a code C as an $[n, k, d]$ code when it is linear where n is the length, k is the dimension, and d is the minimum distance. For non-linear codes, we use the notation (n, M, d) to indicate the same except that $|C| = M$. For codes over rings, this notation is not as useful since we do not have dimension for all rings. While there is a notion of rank, it is not true that two codes of the same rank have the same cardinality and, as such, it is not always an important parameter.

One of the most important algebraic tools for codes is the inner-product. In fact, in many applications, a code is described most naturally as the orthogonal to a code generated by a specific matrix (usually called the parity check matrix). Moreover, this matrix is used to determine if a given vector is in the code and also to decode in many algorithms. We shall define these objects now in our setting.

Let R be a finite commutative ring. To the ambient space R^n, we attach the following Euclidean inner-product:

$$[\mathbf{v}, \mathbf{w}] = \sum_{i=1}^{n} v_i w_i. \tag{1.5}$$

We can now define the standard orthogonal with respect to this inner-product, namely

$$C^\perp = \{\mathbf{v} \mid [\mathbf{v}, \mathbf{w}] = 0, \ \forall \mathbf{w} \in C\}. \tag{1.6}$$

The code $C^\perp$ is linear whether or not C is linear and if C is not linear then $C^\perp = \langle C \rangle^\perp$, where $\langle C \rangle$ denotes the code generated by the vectors of C. It follows immediately from the definition that for linear codes $(C^\perp)^\perp = C$ and for non-linear codes $(C^\perp)^\perp = \langle C \rangle$. Up to this point, we have not used the commutativity of the ring. In order to define $C^\perp$, we need the ring to be commutative. If it were not, then both a left and right orthogonal would need to be defined. Usually, the left and right orthogonals are not equal and in fact need not be of the same size. When the ring is commutative, this difficulty disappears and we have a single orthogonal to a given inner-product.

Example 1.4 Let $C = \langle p \rangle$ be the code of length 1 over $\mathbb{Z}_{pq}$, where p and q are primes. Then $C^\perp = \langle q \rangle$. Here, ideals of the ring can be thought of as linear codes and their annihilator as the dual code.

Example 1.5 The code over any ring given by $C = \langle (1, 1, 1, \ldots, 1) \rangle$ is known as the repetition code. This elementary code was actually used in early NASA missions to correct errors in transmissions. If the ring is $\mathbb{F}_2$ then $C^\perp$ consists of all vectors with evenly many ones. This code is known as the even code E_n of length n.

For some rings and applications, it is preferable to have a Hermitian inner-product using an involution on the ring. In this case, we define the following inner-product:

$$[\mathbf{v}, \mathbf{w}]_H = \sum_{i=1}^{n} v_i \overline{w_i}, \tag{1.7}$$

where $\overline{w_i}$ indicates the involution applied to the element w_i.

For this inner-product we define the orthogonal as

$$C^H = \{\mathbf{v} \mid [\mathbf{v}, \mathbf{w}]_H = 0, \ \forall \mathbf{w} \in C\}. \tag{1.8}$$

As before, C^H is linear whether or not C is linear.

For codes over finite fields, it follows from elementary linear algebra that $\dim(C) + \dim(C^\perp) = n$ and that $\dim(C) + \dim(C^H) = n$. These results will be generalized later in Chap. 3.

We can now give examples of MDS and perfect codes that are linear.

Example 1.6 Let a_i denote n distinct elements of the finite field of order q, $\mathbb{F}_q$. Then let

$$H = \begin{pmatrix} 1 & 1 & 1 & \dots & 1 \\ a_1 & a_2 & a_3 & \dots & a_n \\ a_1^2 & a_2^2 & a_3^2 & \dots & a_n^2 \\ \vdots & \vdots & \vdots & \vdots & \vdots \\ a_1^{d-2} & a_2^{d-2} & a_3^{d-2} & \dots & a_n^{d-2} \end{pmatrix}. \tag{1.9}$$

This matrix is a Vandermonde matrix and has a non-zero determinant. Hence the $d-1$ rows of H are linearly independent and any $d-1$ columns are linearly independent.

Let D_H be the code consisting of the rows of H. Define the code C to be $C = \langle D_H \rangle^{\perp}$. Then C has length n, $|C| = q^{n-(d-1)}$, and has minimum distance d. Then we have that $n - k + 1 = n - (n - (d - 1)) + 1 = d$. Therefore, the code is a linear code which meets the Singleton bound. This gives an infinite family of MDS codes. These codes are known as Reed-Solomon codes and have been widely used in applications, not only because they are optimal, but because there are efficient decoding algorithms associated with them.

Example 1.7 Let $H_{q,r}$ be the r by $\frac{q^r-1}{q-1}$ matrix over $\mathbb{F}_q$ formed by writing the columns as all possible non-zero length r vectors over $\mathbb{F}_q$ and deleting any that are a scalar multiple of a previous vector. This gives that any two columns are linearly independent. Let $C_{q,r} = \langle H_{q,r} \rangle^{\perp}$. Then $C_{q,r}$ has dimension $\frac{q^r-1}{q-1} - r$ and length $n = \frac{q^r-1}{q-1}$. The sphere packing bound gives that the minimum distance is at most 3, and the fact that any two columns of $H_{q,r}$ are linearly independent, gives that the minimum distance is at least 3. Hence $t = 1$. Then $|C_{q,r}|(\sum_{s=0}^{t} C(n, s)(q - 1)^s) = q^{\frac{q^r-1}{q-1}-r}(1 + \frac{q^r-1}{q-1}(q - 1)) = q^{\frac{q^r-1}{q-1}} = q^n$. Therefore the code $C_{q,r}$ is a perfect code. The family of codes $C_{q,r}$ are the well known Hamming codes. If $r = 3$ and $q = 2$, then it is the perfect code given in Example 1.3 constructed from the finite projective plane of order 2.

Other examples of perfect codes are the [23, 12, 7] binary Golay code and the [11, 6, 5] ternary Golay code. The codes were first described in [8].

Notice that the [7, 4, 3] perfect code is not an MDS code as $7 - 4 + 1 = 4 \neq 3$. While any code meeting one of the bounds must be optimal, it does not imply that it will meet the other bound.

Notice that in the examples of linear codes given so far, we have described the code not by giving a generator matrix, but rather by giving the generator matrix of the orthogonal, that is by giving the parity check matrix.

Example 1.8 The ten digit ISBN number for books used a code over $\mathbb{F}_{11}$. The acceptable codewords of length 10 satisfied $\sum_{i=1}^{10} ic_i = 0$. Hence the code of acceptable ISBN numbers was $C = \langle (1, 2, 3, 4, 5, 6, 7, 8, 9, 10) \rangle^{\perp}$. This code was chosen so that it could detect a single error and any double error caused by the transposition of two numbers, these being the most common errors in entering a sequence of numbers.

For codes over a finite field, it is an easy theorem from linear algebra that any code of length n and dimension k is permutation equivalent to a code that has a generator matrix of the form $(I_k \mid A_{n-r})$, where I_k is the k by k identity matrix, and A is a matrix with k rows and $n - k$ columns. It is immediate that if C has such a generator matrix, then $C^\perp$ has a generator matrix of the form $(-A_{n-r}^T \mid I_{n-k})$. This simple fact has a very nice consequence.

Theorem 1.4 *If C is a linear MDS code over a field $\mathbb{F}_q$, then $C^\perp$ is an MDS code.*

Proof The code C is MDS if and only if its parity check matrix $(-A_{n-r}^T \mid I_{n-k})$ has the property that every square submatrix is non-singular. This is because any $n - k$ columns are linearly independent and so the minimum distance of the code must be at least $n - k + 1$ which implies it is MDS. If the matrix has this property, then $(I_k \mid A_{n-r})$, which is the parity check matrix of $C^\perp$, has this property, and so $C^\perp$ is an MDS code as well.

For codes over rings, you cannot always assume that you have a generator matrix of the form $(I_k \mid A_{n-r})$. For example, the code over $\mathbb{Z}_4$, generated by the matrix

$$\begin{pmatrix} 1 & 0 & 1 \\ 0 & 1 & 0 \\ 0 & 0 & 2 \end{pmatrix}, \tag{1.10}$$

has cardinality 32, which is not a power of 4 and hence cannot have a generator matrix of the form $(I_k \mid A_{n-r})$.

In general, the question of a standard form for a generator matrix is more complicated and will be studied in Sect. 2.4.

References

1. Assmus, E.F., Key, J.D.: Designs and their Codes, Cambridge Tracts in Mathematics, 103. Cambridge University Press, Cambridge (1992)
2. Bannai, E., Dougherty, S.T., Harada, M., Oura, M.: Type II codes, even unimodular lattices, and invariant rings. IEEE Trans. Inf. Theory **45**(4), 1194–1205 (1999)
3. Blake, I.F.: Codes over certain rings. Inf. Control **20**, 396–404 (1972)
4. Blake, I.F.: Codes over integer residue rings. Inf. Control **29**(4), 295–300 (1975)
5. Calderbank, A.R., Sloane, N.J.A.: Modular and p-adic cyclic codes. Des. Codes Cryptogr. **6**(1), 21–35 (1995)
6. Delsarte, P.: An algebraic approach to the association schemes of coding theory. Philips Res. Rep. Suppl. **10** (1973)
7. Dougherty, S.T., Park, Y.H.: Codes over the p-adic integers. Des. Codes Cryptogr. **39**(1), 65–80 (2006)
8. Golay, M.J.E.: Notes on digital coding. Proc. IRE **37**, 657 (1949)
9. Hamming, R.W.: Error detecting and error correcting codes. Bell Syst. Tech. J. **29**, 147–160 (1950)
10. Hammons, A.R., Kumar, P.V., Calderbank, A.R., Sloane, N.J.A., Solé, P.: The $\mathbb{Z}_4$-linearity of kerdock, preparata, goethals and related codes. IEEE Trans. Inf. Theory **40**, 301–319 (1994)

11. Hammons, A.R., Kumar, P.V., Calderbank, A.R., Sloane, N.J.A., Solé, P.: On the apparent duality of the Kerdock and Preparata codes. Applied algebra, algebraic algorithms and error-correcting codes (San Juan, PR, 1993). Lecture Notes in Computer Science, vol. 673, pp. 13–24. Springer, Berlin (1993)
12. Huffman, W.C., Pless, V.S.: Fundamentals of Error-Correcting Codes. Cambridge University Press, Cambridge (2003)
13. MacWilliams, F.J., Sloane, N.J.A.: The Theory of Error-Correcting Codes, Amsterdam. North-Holland, The Netherlands (1977)
14. Shannon, C.E.: A mathematical theory of communication. Bell Syst. Techn. J. **27** (1948)
15. Singleton, R.C.: Maximum distance q-nary codes. IEEE Trans. Inf. Theory **10**, 116–118 (1964)
16. Wood, J.: Duality for modules over finite rings and applications to coding theory. Amer. J. Math. **121**(3), 555–575 (1999)

Chapter 2
Ring Theory

In this chapter, we shall give the necessary definitions and foundational results from commutative ring theory for the study of codes over rings.

2.1 Finite Commutative Rings

Rings are one of the fundamental objects of abstract algebra. They have numerous applications in number theory, cryptography, and many other branches of mathematics. For a complete description of ring theory see [12, 14? , 15], and for a description of commutative algebra, see [1].

We shall assume throughout this text that a ring has a multiplicative identity and that the multiplication is commutative. We begin with some standard definitions.

Let R be a finite commutative ring. An ideal $\mathfrak{a}$ of R is an additive subgroup of R such that $r\mathfrak{a} \subseteq \mathfrak{a}$ for all $r \in R$. We note that, in terms of algebraic coding theory, an ideal of R is a code of length 1. An ideal $\mathfrak{m}$ is maximal if $\mathfrak{m}$ is not properly contained in any non-trivial ideal.

Let $\mathfrak{a}$ be an ideal of a finite commutative ring. The chain $\mathfrak{a} \supset \mathfrak{a}^2 \supset \mathfrak{a}^3 \supset \ldots$ necessarily stabilizes. We call the smallest $t \geq 1$ such that $\mathfrak{a}^t = \mathfrak{a}^{t+i}$ for $i \geq 0$ the index of stability of $\mathfrak{a}$. If $\mathfrak{a}$ is nilpotent, then the smallest $t \geq 1$ such that $\mathfrak{a}^t = \{0\}$ is called the index of nilpotency of $\mathfrak{a}$. In this case, it coincides with the index of stability of $\mathfrak{a}$.

Definition 2.1 A ring is a local ring if it has a unique maximal ideal.

Local rings play an important role in coding theory because we often describe rings as the product of local rings via the Chinese Remainder Theorem and reduce much of the theory of codes to the case where the ring is local.

© The Author(s) 2017
S.T. Dougherty, *Algebraic Coding Theory Over Finite Commutative Rings*,
SpringerBriefs in Mathematics, DOI 10.1007/978-3-319-59806-2_2

Example 2.1 Let $R = \mathbb{F}_2[x, y]/\langle x^2, y^2, xy - yx \rangle$. The ring R is a local ring and has cardinality 16. The maximal ideal is $\langle x, y \rangle$. This ring has been widely studied in algebraic coding theory as the ring R_2, see [9].

Definition 2.2 A principal ideal ring is a ring in which each ideal is generated by a single element, that is every ideal $\mathfrak{a}$ can be written as $\mathfrak{a} = \langle a \rangle$ for some element a.

It is well known that $\mathbb{Z}_k$ is a principal ideal ring for all $k > 1$. This family of rings is one of the principal families of rings which are most studied in algebraic coding theory. In fact, they were the first rings which were not fields to be used as alphabets in coding theory, see [2, 3].

Definition 2.3 A chain ring is a principal ideal ring such that the ideals are linearly ordered by set theoretic containment.

It follows that if R is a finite chain ring then there is an element γ such that γ generates the unique maximal ideal and we have the following chain:

$$\{0\} \subseteq \langle \gamma^{e-1} \rangle \subseteq \langle \gamma^{e-2} \rangle \subseteq \cdots \subseteq \langle \gamma \rangle \subseteq R. \tag{2.1}$$

Example 2.2 The ring $\mathbb{Z}_{p^e}$ where p is a prime and $e > 0$ is a chain ring. Here the maximal ideal is $\langle p \rangle$. A Galois ring is a ring of the form $\mathbb{Z}_{p^e}[x]/\langle q(x) \rangle$ where $q(x)$ is irreducible over $\mathbb{Z}_{p^e}$. Galois rings are also chain rings and the maximal ideal is again $\langle p \rangle$.

Let e be the index of nilpotency of the maximal ideal $\langle \gamma \rangle$ of a finite commutative chain ring R. It is shown on page 340 of [15] that for every element a of a chain ring R, we have that there exists a unique integer i with $1 \leq i \leq e - 1$ such that $a = \mu \gamma^i$, with μ a unit.

It follows that a chain ring is necessarily a local ring, but a local ring need not be a chain ring, as in the following example.

Example 2.3 Let $R = \mathbb{Z}_4[x]/\langle x^2 \rangle$. Then R is a ring of order 16 with maximal ideal $\langle 2, x \rangle = \{0, x, 2x, 3x, 2, 2 + x, 2 + 2x, 2 + 3x\}$. But the ideals $\langle 2 \rangle = \{0, 2, 2x, 2 + 2x\}$ and $\langle x \rangle = \{0, x, 2x, 3x\}$ are not linearly ordered.

Let R be a commutative chain ring with maximal ideal $\langle \gamma \rangle$ with index of nilpotency e. We know that $R/\langle \gamma \rangle$ is isomorphic to a finite field $\mathbb{F}_{p^r}$. It is well known that $|\langle \gamma^j \rangle| = |\mathbb{F}_{p^r}|^{e-j}$ for $0 \leq j \leq e - 1$. It follows that

$$|R| = |\mathbb{F}_{p^r}||\langle \gamma \rangle| = |\mathbb{F}_{p^r}||\mathbb{F}_{p^r}|^{e-1} = p^{er}. \tag{2.2}$$

We give the next definition in terms of commutative rings. This definition would be slightly changed in the case of non-commutative rings. For a complete description of the Jacobson radical and socle in their general usage see [12?].

Definition 2.4 Let R be a commutative ring. Then the Jacobson radical $J(R)$ of a ring R can be characterized as the intersection of all maximal ideals.

In any ring, commutative or non-commutative, the Jacobson radical is a two sided ideal. In a local commutative ring, the Jacobson radical is necessarily the unique maximal ideal.

Definition 2.5 The nilradical of a commutative ring R consists of the nilpotent elements of the ring.

As we shall prove later, for finite commutative rings, the Jacobson radical and the nilpotent radical coincide. Like the definition of the Jacobson radical, the following definition of the socle of the ring would change slightly for rings in general.

Definition 2.6 Let R be a commutative ring. The socle of a ring R, $Soc(R)$, is defined as the sum of all the minimal ideals of the ring.

2.2 Frobenius Rings

For algebraic coding theory, the most important class of rings is the class of Frobenius rings. This is because both MacWilliams theorems apply for Frobenius rings, but, in general, they do not extend to larger families of rings, see Theorems 2.5 and 3.2. In essence what this means is that for codes over this class of rings we have many of the techniques and ideas that fuel classical coding theory over fields. In spaces where these two theorems do not hold, things act in a very different way than classical coding theory and we lose much of the power of the theory. Perhaps one of the most significant implications of this is that for a code C over a Frobenius ring R of length n, we have that $|C||C^{\perp}| = |R^n|$. This is not necessarily true when the ring is not Frobenius. These results were first introduced in [17, 18].

In this section, we shall give a very brief explanation of Frobenius rings. Our explanation is based on Nakayama's definition. However, we shall not discuss Frobenius rings in their broadest generality, but rather reduce definitions to their finite commutative case. For example, one would generally begin the discussion with left (right) Artinian rings, namely those that do not contain an infinite descending chain of left (right) ideals. Since we only consider finite rings, all of these rings are Artinian and we need not consider ideals as being left or right, since all ideals are two sided ideals in a commutative ring. When we want to stress that something is a module or an ideal in a ring, we shall use the notation as a left module or ideal. For a more general description of Frobenius rings as applied to coding theory, including the non-commutative case, see [5].

Recall that a module M is irreducible if it contains no non-trivial submodule and a module M is indecomposable if it has no non-trivial direct summands. We note that every irreducible module is indecomposable, but not the converse.

Any Artinian ring, as a module over itself, admits a finite direct sum decomposition, namely:

$$_R R = Re_{1,1} \oplus \ldots Re_{1,\mu_1} \oplus \cdots \oplus Re_{n,1} \oplus \cdots \oplus Re_{n,\mu_n}, \tag{2.3}$$

where the $e_{i,j}$ are primitive orthogonal idempotents with $1 = \sum e_{i,j}$. This decomposition is known as the principal decomposition of the module of R over itself.

We index the $Re_{i,j}$ so that $Re_{i,j}$ is isomorphic to $Re_{k,l}$ if and only if $i = k$. Then we set $e_i = e_{i,1}$ and we write $_R R \cong \oplus \mu_i Re_i$.

We can extend the definition of socle and radical of a ring to a module in a natural way. That is, the socle of a module M is the sum of the simple (i.e. contains no non-zero submodules) submodules of M and the radical of a module M is the intersection of all maximal submodules of M. Then the module $Re_{i,j}$ has a unique maximal submodule $Rad(R)e_{i,j} = Re_{i,j} \cap Rad(R)$ and a unique irreducible top quotient $T(Re_{i,j}) = Re_{i,j}/Rad(R)e_{i,j}$. The socle $S(Re_{,j})$ is the submodule generated by the irreducible submodules of $Re_{i,j}$.

We can now proceed to the standard definition. Let the module of R over itself be decomposed as follows: $_R R = \oplus \mu_i Re_i$. Then, an Artinian ring R is quasi-Frobenius if there exists a permutation σ of $\{1, 2, \ldots, n\}$, such that

$$T(Re_i) \cong S(Re_{\sigma(i)}) \tag{2.4}$$

and

$$S(Re_i) \cong T(Re_{\sigma(i)}). \tag{2.5}$$

Then the ring is Frobenius if $\mu_{\sigma(i)} = \mu_i$ as well.

A module M over a ring R is injective if, for every pair of R-modules $B_1 \subset B_2$ and every R-linear mapping $f : B_1 \to M$, the mapping f extends to an R-linear mapping $\overline{f} : B_2 \to M$.

The proof of the following can be found in Theorem 1.2 and Remark 1.3 of [17].

Theorem 2.1 *Let R be a finite commutative ring, then the following conditions are equivalent:*

- *The ring R is Frobenius;*
- *the R-module R is injective.*
- *If R is a finite local ring with maximal ideal $\mathfrak{m}$ and residue field $\mathbf{k}$, these conditions are equivalent with $\dim_{\mathbf{k}} Ann(\mathfrak{m}) = 1$.*

Example 2.4 Consider the ring $R = \mathbb{F}_2[x, y]/\langle x^2, y^2, xy \rangle$. We have that $|R| = 8$ and R has a maximal ideal $\mathfrak{m} = \{0, x, y, x + y\}$. Notice that $\mathfrak{m} = \mathfrak{m}^\perp$. Then $\dim_{\mathbf{k}} Ann(\mathfrak{m}) = 2$ which violates the last condition in Theorem 2.1. Hence R is not Frobenius. In this case, we have that $|\mathfrak{m}||\mathfrak{m}^\perp| \neq |R|$. In a Frobenius ring this is not possible.

Throughout this text, we view characters as homomorphisms $\chi : M \to \mathbb{C}^*$ rather than maps into $\mathbb{Q}/\mathbb{Z}$. For a module M, let $\widehat{M}$ denote the character module of M. One of the most important aspects of Frobenius rings in terms of algebraic coding theory is the characterization of their character module. The following theorem can be found in [17]. It characterizes Frobenius rings in terms of the character module.

Theorem 2.2 *Suppose R is a finite ring. The following are equivalent:*

- *The ring R is Frobenius.*
- *As a left module, $\widehat{R} \cong {_R}R$.*
- *As a right module $\widehat{R} \cong R_R$.*

Note that the result is more complex in terms of non-commutative rings since we must be concerned with whether the module is a left or a right module.

Let R be a Frobenius ring. Let $\phi : R \to \widehat{R}$ be the module isomorphism. Then set $\chi = \phi(1)$ so that $\phi(r) = \chi^r$ for $r \in R$. We call this character χ a generating character for $\widehat{R}$.

The following is an immediate consequence.

Theorem 2.3 *The finite commutative ring R is Frobenius if and only if $\widehat{R}$ has a generating character.*

Example 2.5 Consider the finite field $\mathbb{F}_p$ where p is a prime. Let ξ be a complex primitive p-th root of unity. Then $\chi(a) = \xi^a$ is a generating character for $\widehat{\mathbb{F}_p}$.

The generating character for a Frobenius ring R is not necessarily unique. In fact, we have the following theorem, which is Lemma 4.1 in [17], where it is stated in broader generality for the non-commutative case as well.

Theorem 2.4 *Let χ be a character of a finite commutative ring R. Then χ is a generating character if and only if $\ker(\chi)$ contains no nonzero ideals of R.*

Example 2.6 Consider the finite field $\mathbb{F}_4$ where the elements are written as $a + b\omega$ for $a, b \in \mathbb{F}_2$. Then the character $\chi_1 : \mathbb{F}_4 \to \mathbb{C}$ defined by $\chi_1(a + b\omega) = (-1)^{a+b}$ is a generating character for $\widehat{\mathbb{F}_4}$. Additionally, the character $\chi_2 : \mathbb{F}_4 \to \mathbb{C}$ defined by $\chi_2(a + b\omega) = (-1)^b$ is a generating character for $\widehat{\mathbb{F}_4}$. Their respective character tables are given by the following, where the value for row α and column β is $\chi_i(\alpha\beta)$.

χ_1	0	1	ω	$1+\omega$		χ_2	0	1	ω	$1+\omega$
0	1	1	1	1		0	1	1	1	1
1	1	-1	-1	1		1	1	1	-1	-1
ω	1	-1	1	-1		ω	1	-1	-1	1
$1+\omega$	1	1	-1	-1		$1+\omega$	1	-1	1	-1

The tables described in the previous example are very important in terms of coding theory since they will be used to produce MacWilliams relations for codes over rings. See Chap. 3 for a full description.

The final characterization of Frobenius rings that we shall give is the following extension of MacWilliams' first theorem, which she had proven for finite fields. It was extended in [17] to the following theorem. In [?], it was shown that this theorem does not extend to quasi-Frobenius rings. We state the theorem here without proof. A detailed proof can be found in [17].

Theorem 2.5 (MacWilliams Theorem) *(A) If R is a finite Frobenius ring and C is a linear code over R, then every Hamming isometry $C \to R^n$ can be extended to a monomial transformation.*

(B) If a finite commutative ring R satisfies that all of its Hamming isometries between linear codes allow for monomial extensions, then R is a Frobenius ring.

This theorem, along with the MacWilliams relations in Chap. 3, explain why we use Frobenius rings as the alphabets for codes. Specifically, we want both of these theorems to be true in order to apply the most powerful results of algebraic coding theory to codes over rings.

2.3   Chinese Remainder Theorem

The most powerful tool for codes over commutative rings is the classical Chinese Remainder Theorem, which we now describe. For a full description of the approach to the Chinese Remainder Theorem see [14].

Definition 2.7 Two ideals $\mathfrak{a}$ and $\mathfrak{b}$ of a ring R are said to be relatively prime if $\mathfrak{a} + \mathfrak{b} = R$.

Occasionally, the term coprime is used instead of relatively prime for ideals satisfying this definition.

Lemma 2.1 *If $\mathfrak{a}$ and $\mathfrak{b}$ are relatively prime ideals of a commutative ring R, then $\mathfrak{ab} = \mathfrak{a} \cap \mathfrak{b}$.*

Proof It is immediate that $\mathfrak{ab} \subseteq \mathfrak{a} \cap \mathfrak{b}$. If $\mathfrak{a} + \mathfrak{b} = R$, then $\mathfrak{a} \cap \mathfrak{b} = (\mathfrak{a} \cap \mathfrak{b})R = (\mathfrak{a} \cap \mathfrak{b})(\mathfrak{a} + \mathfrak{b}) \subseteq \mathfrak{ab}$. Therefore $\mathfrak{ab} = \mathfrak{a} \cap \mathfrak{b}$.   □

Lemma 2.2 *Let $\mathfrak{a}$, $\mathfrak{b}$ and $\mathfrak{c}$ be ideals of a commutative ring R that are relatively prime in pairs. Then $\mathfrak{a}$ is relatively prime to $\mathfrak{bc}$.*

Proof We have that $R = (\mathfrak{a} + \mathfrak{b})(\mathfrak{a} + \mathfrak{c}) \subseteq \mathfrak{a} + \mathfrak{bc}$. Therefore $\mathfrak{a} + \mathfrak{bc} = R$ and $\mathfrak{a}$ and $\mathfrak{bc}$ are relatively prime.   □

Apply Lemmas 2.1 and 2.2 inductively and we have the following.

Lemma 2.3 *Let $\mathfrak{a}_1, \mathfrak{a}_2, \ldots, \mathfrak{a}_s$ be ideals of a commutative ring R that are relatively prime in pairs. Then $\mathfrak{a}_1 \mathfrak{a}_2 \ldots \mathfrak{a}_s = \mathfrak{a}_1 \cap \mathfrak{a}_2 \cap \cdots \cap \mathfrak{a}_s$.*

Next we can use this to produce an isomorphism lemma.

Lemma 2.4 *Let $\mathfrak{a}$ and $\mathfrak{b}$ be relatively prime ideals of a commutative ring R. Then $R/\mathfrak{ab} \cong R/\mathfrak{a} \times R/\mathfrak{b}$.*

Proof Define the map $\Psi : R \to (R/\mathfrak{a} \times R/\mathfrak{b})$ by $\Psi(x) = (x \pmod{\mathfrak{a}}, x \pmod{\mathfrak{b}})$. We have $ker(\Psi) = \mathfrak{a} \cap \mathfrak{b} = \mathfrak{ab}$, which gives that $R/\mathfrak{ab} \cong R/\mathfrak{a} \times R/\mathfrak{b}$.   □

Computationally we have the following. Since $\mathfrak{a} + \mathfrak{b} = R$, there exists an $\alpha \in \mathfrak{a}$ and a $\beta \in \mathfrak{b}$ with $\alpha + \beta = 1$. Then $\Psi(c\alpha + d\beta) = (d, c)$. Specifically, Ψ is surjective and we can compute the preimage in a straightforward computation. Applying induction to Lemma 2.4 we have the following.

Lemma 2.5 *Let* $\mathfrak{a}_1, \mathfrak{a}_2, \ldots, \mathfrak{a}_s$ *be ideals of a commutative ring R which are relatively prime in pairs. Then*

$$R/\mathfrak{a}_1\mathfrak{a}_2\ldots\mathfrak{a}_s \cong R/\mathfrak{a}_1 \times R/\mathfrak{a}_2 \times \cdots \times R/\mathfrak{a}_s. \tag{2.6}$$

Let R be a finite commutative ring, with $\mathfrak{a}$ an ideal of R. Let $\Psi_\mathfrak{a}$ be the canonical homomorphism $\Psi_\mathfrak{a} : R \to R/\mathfrak{a}$, given by $\Psi_\mathfrak{a}(x) = x + \mathfrak{a}$.

Let $\mathfrak{m}_1, \ldots, \mathfrak{m}_s$ be the maximal ideals of a finite commutative ring R and let $e_1, \ldots, e_s$ be their respective indices of stability. The ideals $\mathfrak{m}_1^{e_1}, \ldots, \mathfrak{m}_s^{e_s}$ are relatively prime in pairs and $\prod_{i=1}^{s} \mathfrak{m}_i^{e_i} = \cap_{i=1}^{k} \mathfrak{m}_i^{e_i} = \{0\}$.

This leads us to the following well known theorem.

Theorem 2.6 (Chinese Remainder Theorem) *Let R be a finite commutative ring, with maximal ideals $\mathfrak{m}_1, \ldots, \mathfrak{m}_s$ where the index of stability of $\mathfrak{m}_i$ is e_i. Then the map $\Psi : R \to \prod_{i=1}^{s} R/\mathfrak{m}_i^{e_i}$, defined by $\Psi(x) = (x + \mathfrak{m}_1^{e_1}, \ldots, x + \mathfrak{m}_k^{e_k})$, is a ring isomorphism.*

Proof We have that the $\mathfrak{m}^{e_i}$ are relatively prime in pairs and $\cap_{i=1}^{s} m_i^{e_i} = \{0\}$. Then by Lemma 2.5 we have that $R \cong R/0 \cong R/\mathfrak{m}_1^{e_1} \times R/\mathfrak{m}_2^{e_2} \times \cdots \times R/\mathfrak{m}_k^{e_k}$. This gives the result.   □

Let R_i denote the local ring $R/\mathfrak{m}_i^{e_i}$. The previous theorem gives that

$$R \cong R_1 \times R_2 \times \cdots \times R_s. \tag{2.7}$$

We note that R is Frobenius if and only if each R_i is Frobenius. See Remark 1.3 in [17] for an explanation.

We denote the inverse isomorphism of Ψ by CRT, so that $CRT : R_1 \times R_2 \times \cdots \times R_s \to R$.

Example 2.7 Let $\prod_{i=1}^{s} \cdot p_i^{e_i}$ be the prime factorization of a positive natural number n. Then by Theorem 2.6 we have that $\mathbb{Z}_n \cong \mathbb{Z}_{p_1^{e_1}} \times \mathbb{Z}_{p_2^{e_2}} \times \cdots \times \mathbb{Z}_{p_s^{e_s}}$. This is the classical application of the Chinese Remainder Theorem and is where the name originates. Namely, it allows for the unique solution modulo $\prod n_i$ of the system of equations $x \equiv a_i \pmod{n_i}$ when the n_i are relatively prime in pairs.

By an abuse of notation, we extend both Ψ and CRT to the n fold product of their domains.

If C_i is a code over R_i, we let $C = CRT(C_1, C_2, \ldots, C_s)$ be the code over R formed by this extended isomorphism. It is immediate that any code C over R is the image of a some collection of codes $C_1, C_2, \ldots, C_s$ where C_i is a code over R_i.

The rank of a code, $rank(C)$, is the minimum number of generators of C. A code is said to be free if it is a free submodule over R. The following appears in [7].

Corollary 2.1 *Let R_i be finite commutative rings and let*

$$R = CRT(R_1, R_2, \ldots, R_s).$$

Let C_i be a code over R_i with $C = CRT(C_1, C_2, \ldots, C_s)$. Then

- $|C| = \prod_{i=1}^{s} |C_i|$;
- $rank(C) = max\{rank(C_i), i = 1, \ldots, s\}$;
- *C is free if and only if C_i is free for all i each of the same rank.*

Proof The first statement follows immediately from the fact that CRT is a bijection. To prove the second, let r_i be the rank of C_i and $\mathbf{v}_1^i, \mathbf{v}_2^i, \ldots, \mathbf{v}_{r_i}^i$ be a set of generators for C_i. Let r be the maximum value for r_i. Pad this set with zero vectors so that each generator set has r elements. Then

$$\{CRT(\mathbf{v}_1^1, \mathbf{v}_1^2, \ldots, \mathbf{v}_1^s), CRT(\mathbf{v}_2^1, \mathbf{v}_2^2, \ldots, \mathbf{v}_2^s), \ldots, CRT(\mathbf{v}_r^1, \mathbf{v}_r^2, \ldots, \mathbf{v}_r^s)\}$$

generates the code C. We need r vectors since there exists an i where $r_i = r$. It follows from this construction that the code C is free if and only if $r_i = r$ for each i and each code C_i is free. □

The following is a well known application of the Chinese Remainder Theorem.

Theorem 2.7 *Let $R = CRT(R_1, R_2, \ldots, R_s)$ be a finite commutative ring. Let $C = CRT(C_1, C_2, \ldots, C_s)$ be a code over R. Then*

$$C^\perp = CRT(C_1^\perp, C_2^\perp, \ldots, C_s^\perp). \tag{2.8}$$

Proof Consider vectors $\mathbf{v}, \mathbf{w} \in R^n$. Then $\Psi_\mathfrak{a}(\sum v_i w_i) = \sum \Psi_\mathfrak{a}(v_i) \sum \Psi_\mathfrak{a}(w_i)$. Hence, when $[\mathbf{v}, \mathbf{w}] = 0$, we have that $[\Psi_\mathfrak{a}(\mathbf{v}), \Psi_\mathfrak{a}(\mathbf{w})] = 0$. Then the standard cardinality argument gives equality. □

A similar proof gives the following theorem.

Theorem 2.8 *Let $R = CRT(R_1, R_2, \ldots, R_s)$ be a finite commutative ring. Let $C = CRT(C_1, C_2, \ldots, C_s)$ be a code over R. If $\overline{\mathfrak{a}} = \mathfrak{a}$ and $\Psi_\mathfrak{a}(\overline{\mathbf{v}}) = \overline{\Psi_\mathfrak{a}(\mathbf{v})}$, where the involution applies first in the ring R and then in the ring $R/\mathfrak{a}$, then $C^H = CRT(C_1^H, C_2^H, \ldots, C_s^H)$.*

We can also find the minimum weight of a code in terms of its components via the Chinese Remainder Theorem as in the following theorem.

Theorem 2.9 *Let $R = CRT(R_1, R_2, \ldots, R_s)$ be a finite commutative ring. Let $C = CRT(C_1, C_2, \ldots, C_s)$ be a code over R. Then $d(C) = min\{d(C_i)\}$.*

Proof Let d_1 be the minimum of $\{d(C_i)\}$. Then, there exists j with $d(C_j) = d_1$. Let $\mathbf{v}_j$ be a minimum weight vector in C_j, then

$$CRT(\mathbf{0}, \mathbf{0}, \ldots, \mathbf{0}, \mathbf{v}_j, \mathbf{0}, \ldots, \mathbf{0})$$

has Hamming weight d_1 which gives $d(C) \leq d_1$. Then let $\mathbf{v}$ be a minimum weight vector in C. Its projection $\Psi_\alpha(\mathbf{v})$ has weight less than or equal to $d(C)$ which gives $d(C) \geq d_1$. Therefore, $d_1 = d(C)$, and we have the result. $\qquad\square$

Recall that an ideal $\mathfrak{a}$ is prime if $ab \in \mathfrak{a}$ implies either $a \in \mathfrak{a}$ of $b \in \mathfrak{a}$. In a finite ring, prime ideals and maximal ideals coincide since finite division rings are fields. Therefore the nilradical and the Jacobson radical coincide. Moreover, since the ring is finite, the nilradical is nilpotent. This is because you can simply take the maximum nilpotency exponent of all nilpotent elements and apply this to the ideal.

Theorem 2.10 *Let R be a finite commutative ring. Then R is isomorphic, via the Chinese Remainder Theorem, to a direct product of local rings.*

Proof Let $\mathfrak{m}_1, \mathfrak{m}_2, \ldots, \mathfrak{m}_s$ be the maximal ideals of R. Then the Jacobson radical $J(R) = \mathfrak{m}_1 \cap \mathfrak{m}_2 \cap \cdots \cap \mathfrak{m}_s = \mathfrak{m}_1 \mathfrak{m}_2 \ldots \mathfrak{m}_s$ by Lemma 2.3. Since $J(R)$ is nilpotent we have that there exists k with $(J(R)^k) = \{0\}$. This gives that $(\mathfrak{m}_1 \mathfrak{m}_2 \ldots \mathfrak{m}_s)^k = \mathfrak{m}_1^k \mathfrak{m}_2^k \ldots \mathfrak{m}_s^k = \{0\}$. We know that $\mathfrak{m}_i$ and $\mathfrak{m}_j$ are relatively prime for $i \neq j$. Then their powers are also relatively prime by Lemma 2.2. This allows us to invoke the Chinese Remainder Theorem, which gives us that R is isomorphic to $R/\mathfrak{m}_1^k \times R/\mathfrak{m}_2^k \times \cdots \times R/\mathfrak{m}_s^k$. (Notice that k is greater than or equal to the individual index of stabilities of the maximal ideals so that $\mathfrak{m}_i^k = \mathfrak{m}_i^{e_i}$.) It only remains to show that $R/\mathfrak{m}_i^k$ is local. A maximal ideal in $R/\mathfrak{m}_i^k$ corresponds to a maximal ideal $\mathfrak{a}$ of R with $\mathfrak{m}_i \subseteq \mathfrak{a}$ since $\mathfrak{a}$ is necessarily a prime ideal. Then since $\mathfrak{m}_i$ is maximal, we have that $\mathfrak{a} = \mathfrak{m}_i$. Therefore, the unique maximal ideal of $R/\mathfrak{m}_i$ is $\mathfrak{m}_i/\mathfrak{m}_i^k$. $\qquad\square$

Theorem 2.11 *A finite commutative ring R is a principal ideal ring if and only if $R = CRT(R_1, R_2, \ldots, R_s)$ where R_i is a chain ring for all i.*

Proof Assume $R = R_1 \times R_2 \times \cdots \times R_s$ and each R_i is a chain ring. Chain rings are necessarily principal ideal rings. If $\mathfrak{a}_i$ is an ideal of R_i with $\mathfrak{a}_i = \langle a_i \rangle$ then the ideal $\mathfrak{a}_1 \times \mathfrak{a}_2 \times \cdots \times \mathfrak{a}_s$ in $R_1 \times R_2 \times \cdots \times R_s$ is principal and generated by $(a_1, a_2, \ldots, a_s)$. Hence R is principal.

Assume R is principal. Then any ideal in $R_1 \times R_2 \times \cdots \times R_s$ is principal and hence each R_i is principal. By Theorem 2.10 we have that each R_i is local. Therefore R_i is a principal ideal ring which is local and hence a chain ring. $\qquad\square$

The standard example of this theorem is the example given in Example 2.7. Namely, $\mathbb{Z}_n \cong \mathbb{Z}_{p_1^{e_1}} \times \mathbb{Z}_{p_2^{e_2}} \times \cdots \times \mathbb{Z}_{p_s^{e_s}}$. Here $\mathbb{Z}_n$ is a principal ideal ring and each $\mathbb{Z}_{p_i^{e_i}}$ is a chain ring.

Example 2.8 For integers $k \geq 1$, define the family of rings A_k to be $A_k = \mathbb{F}_2[v_1, v_2, \ldots, v_k]/\langle v_i^2 - v_i, v_i v_j - v_j v_i\rangle$. The ideal $\langle w_1, w_2, \ldots, w_k\rangle$, where $w_i \in \{v_i, 1 + v_i\}$, is a maximal ideal of cardinality 2^{2^k-1}. We denote these maximal ideals by $\mathfrak{m}_i$. We note that here are 2^k such ideals and that $\mathfrak{m}_i^e = \mathfrak{m}_i$ for all i and $e \geq 1$. It is elementary to see that the direct sum of any two of these ideals is A_k. Then, using the Chinese Remainder Theorem, we have that the ring A_k is isomorphic to $\mathbb{F}_2^{2^k}$. As such, the ring A_k is a principal ideal ring and is isomorphic to the direct product of chain rings. Codes over these rings were studied in [4].

2.4   Generators

One of the most important tools in coding theory is finding a generator matrix for a code. In general, we do not only want a matrix whose rows generate the code, but we want a matrix that generates the code with the minimum number of rows. For codes over fields, we have a simple determination of a minimal generating set. Namely, a set of vectors $\mathbf{v}_1, \mathbf{v}_2, \ldots, \mathbf{v}_n$ is linearly independent if $\sum \alpha_i \mathbf{v}_i = \mathbf{0}$ implies $\alpha_i = 0$ for all i. This standard definition and its implications from linear algebra gives that any code over a finite field is equivalent to a code that has a minimal generating matrix of the form $(I_k \mid A)$ where I_k is the k by k identity matrix. For codes over rings this is not always possible. For example, the code of length 1 over $\mathbb{Z}_4$ generated by 2 is the code $\{0, 2\}$. This code has no such matrix. Moreover, the minimality of a set of generators can also be quite different. For example, consider the code C over $\mathbb{Z}_6$ of length 2 generated by the following matrix:

$$\begin{pmatrix} 2 & 0 \\ 0 & 3 \end{pmatrix}.$$

Here we have that $|C| = 6$ and it may appear that this generating set is minimal, however, the vector $(2, 3)$ also generates the code which shows that the original set of generators was not minimal. In this section, we shall describe the theory for minimal generating sets for codes over rings. Much of this material was first presented in [8, 16].

Definition 2.8 Let R be a finite local commutative Frobenius ring with unique maximal ideal $\mathfrak{m}$, and let $\mathbf{v}_1, \mathbf{v}_2, \ldots, \mathbf{v}_s$ be vectors in R^n. Then $\mathbf{v}_1, \mathbf{v}_2, \ldots, \mathbf{v}_s$ are modular independent if and only if $\sum \alpha_j \mathbf{v}_j = \mathbf{0}$ implies that $\alpha_j \in \mathfrak{m}$ for all j.

A finite field is a local ring with maximal ideal $\{0\}$, so this definition is a natural generalization of linear independence. As an example for a code over a ring, consider the generators $(2, 0), (0, 4)$ over the local ring $\mathbb{Z}_8$. These vectors are modular independent since any linear combination summing to the zero vector implies that the coefficients are in the maximal ideal $\langle 2 \rangle$. The following lemma is a natural generalization for one of the primary implications of linear independence.

Lemma 2.6 *Let R be a finite local commutative Frobenius ring and let* $\mathbf{v}_1, \mathbf{v}_2, \ldots, \mathbf{v}_s \in R^n$. *Then* $\mathbf{v}_1, \mathbf{v}_2, \ldots, \mathbf{v}_s$ *are modular dependent if and only if some* $\mathbf{v}_j$ *can be written as a linear combination of the other vectors.*

Proof Assume that the vectors $\mathbf{v}_1, \mathbf{v}_2, \ldots, \mathbf{v}_s$ are modular dependent. This implies that there exists α_i with $\sum \alpha_i \mathbf{v}_i = \mathbf{0}$ and there exists α_j such that $\alpha_j \notin \mathfrak{m}$. We have that $\mathfrak{m}$ must contain all non-units giving that α_j is a unit. Then we have that

$$\mathbf{v}_j = (-\alpha_j^{-1}\alpha_1)\mathbf{v}_1 + \cdots + (-\alpha_j^{-1}\alpha_{j-1})\mathbf{v}_{j-1} + \\ (-\alpha_j^{-1}\alpha_{j+1})\mathbf{v}_{j+1} + \cdots + (-\alpha_j^{-1}\alpha_s)\mathbf{v}_s.$$

To prove the other direction, assume that $\mathbf{v}_j$ can be written as a linear combination of the other vectors. Then $\mathbf{v}_j = \sum_{i \neq j}^{s} \alpha_i \mathbf{v}_i$. Then we have that $\sum_{i \neq j}^{s} \alpha_i \mathbf{v}_i - \mathbf{v}_j = \mathbf{0}$. The coefficient -1 is a unit in R and this implies that $\mathbf{v}_1, \mathbf{v}_2, \ldots, \mathbf{v}_s$ are modular dependent. $\square$

In terms of finite local commutative Frobenius rings, this result is enough to determine minimal generating sets. Namely, we need a set of modular independent vectors. For chain rings, we can say more. Since the ideals are in a chain we can apply the previous lemma and the standard techniques of linear algebra (that is row reduction done over a chain ring) to obtain the following result.

Theorem 2.12 *Let R be a finite chain ring with maximal ideal* $\langle \gamma \rangle$ *and let C be a code over R. Then there exists a generator matrix for a code C over R that is permutation equivalent to the following:*

$$\begin{pmatrix} I_{k_0} & A_{0,1} & A_{0,2} & A_{0,3} & \cdots & & & A_{0,e} \\ 0 & \gamma I_{k_1} & \gamma A_{1,2} & \gamma A_{1,3} & \cdots & & & \gamma A_{1,e} \\ 0 & 0 & \gamma^2 I_{k_2} & \gamma^2 A_{2,3} & \cdots & & & \gamma^2 A_{2,e} \\ \vdots & \vdots & 0 & \ddots & \ddots & & & \vdots \\ \vdots & \vdots & \vdots & \ddots & \ddots & \ddots & & \vdots \\ 0 & 0 & 0 & \cdots & 0 & \gamma^{e-1} I_{k_{e-1}} & \gamma^{e-1} A_{e-1,e} \end{pmatrix}, \quad (2.9)$$

where the $A_{i,j}$ *are arbitrary matrices with elements from the ring R and* I_{k_j} *is the* k_j *by* k_j *identity matrix.*

A code with generator matrix of this form is said to have type $\{k_0, k_1, \ldots, k_{e-1}\}$. The following is an immediate consequence of Theorem 2.12.

Corollary 2.2 *Let R be a finite chain ring with maximal ideal* $\langle \gamma \rangle$. *Let C be a code over R of type* $\{k_0, k_1, \ldots, k_{e-1}\}$. *Then,*

$$|C| = |R/\langle \gamma \rangle|^{\sum_{i=0}^{e-1}(e-i)k_i}. \quad (2.10)$$

For a code over a finite chain ring the type plays the role that the dimension plays for codes over a field. This is because two codes with the same type will have the same cardinality. This is not true for two codes with the same rank as a module over the ring.

We now expand this theory to cover any finite commutative Frobenius ring.

Definition 2.9 Let R be a finite commutative Frobenius ring with

$$R = CRT(\Psi_1(R), \Psi_2(R), \ldots, \Psi_s(R)) = (R_1, R_2, \ldots, R_s).$$

The vectors $\mathbf{v}_1, \cdots, \mathbf{v}_k$ in R^n are modular independent if $\Psi_i(\mathbf{v}_1), \cdots, \Psi_i(\mathbf{v}_k)$ are modular independent for some i, with $1 \leq i \leq s$.

Note that we are only requiring that their image under Ψ_i be modular independent over one local ring. They need not be modular independent for all i.

Theorem 2.13 *Let R be a finite commutative Frobenius ring and let $\mathbf{v}_1, \cdots, \mathbf{v}_k$ be vectors that are modular independent over R. If $\sum \alpha_j \mathbf{v}_j = \mathbf{0}$, then α_j is not a unit for all j.*

Proof Let $R = CRT(\Psi_1(R), \Psi_2(R), \ldots, \Psi_s(R))$ and let i be the index such that $\Psi_i(\mathbf{v}_1), \cdots, \Psi_i(\mathbf{v}_k)$ are modular independent over the local ring R_i. Then $\sum \alpha_j \mathbf{v}_j = \mathbf{0}$ implies that $\sum \Psi_i(\alpha_j)\Psi_i(\mathbf{v}_j) = \mathbf{0}$, and hence we have that $\Psi_i(\alpha_j) \in \mathfrak{m}_i$ where $\mathfrak{m}_i$ is the maximal ideal of R_i.

If α_j were a unit in R then there would exist a $\beta \in R$ with $\alpha_j \beta = 1$, which gives that $\Psi_i(\alpha_j)\Psi_i(\beta) = 1$. This would imply that $\Psi_i(\alpha_j)$ is a unit in R_i, which would be a contradiction. Therefore, we have that $\Psi_i(\alpha_j)$ is not a unit for all j. $\square$

The converse of this theorem is not true. For example, consider the vectors $(5, 5)$ and $(7, 7)$ over $\mathbb{Z}_{35}$. If $\alpha_1(5, 5) + \alpha_2(7, 7) = (0, 0)$, then α_1 and α_2 must be non-units. However, these vectors are not modular independent over $\mathbb{Z}_{35}$, since their images under Ψ_1 and Ψ_2 are not modular independent over $\mathbb{Z}_5$ or $\mathbb{Z}_7$.

Because we do not have the biconditional yet, we need something else in the case when the ring is not local.

Definition 2.10 Let R be a finite commutative Frobenius ring. Let $\mathbf{v}_1, \cdots, \mathbf{v}_k$ be non-zero vectors in R^n. Then $\mathbf{v}_1, \cdots, \mathbf{v}_k$ are independent if $\sum \alpha_j \mathbf{v}_j = \mathbf{0}$ implies that $\alpha_j \mathbf{v}_j = \mathbf{0}$ for all j.

Note that we are saying something different than the coefficient is zero. We are saying that the vector $\alpha_j \mathbf{v}_j = \mathbf{0}$. Again, for a code over a field, this would imply that the coefficient is 0 since we have no zero divisors. This definition would then reduce to the standard definition for linear independence for vectors over a field.

Theorem 2.14 *Let R be a finite commutative Frobenius ring with $\mathbf{v}_1, \cdots, \mathbf{v}_k$ vectors over R. If $\mathbf{v}_1, \cdots, \mathbf{v}_k$ are independent and $\alpha \mathbf{w} \notin \langle \mathbf{v}_1, \cdots, \mathbf{v}_k \rangle$, for all $\alpha \neq 0$, then the vectors $\mathbf{v}_1, \cdots, \mathbf{v}_k, \mathbf{w}$ are independent.*

Proof If $\sum \alpha_j \mathbf{v}_j + \beta \mathbf{w} = \mathbf{0}$, then $\sum \alpha_j \mathbf{v}_j = -\beta \mathbf{w}$, which is a contradiction since then $-\beta \mathbf{w}$ would be in the span of the $\mathbf{v}_i$, unless $\beta = 0$. If $\beta = 0$ then $\sum \alpha_j \mathbf{v}_j = \mathbf{0}$, and then $\alpha_j \mathbf{v}_j = \mathbf{0}$ for all j since $\mathbf{v}_1, \cdots, \mathbf{v}_k$ are independent. Therefore, we have that the vectors $\mathbf{v}_1, \cdots, \mathbf{v}_k, \mathbf{w}$ are independent. $\square$

Following Definition 2.10, we can easily get the following theorem.

Theorem 2.15 *Let R be a finite commutative Frobenius ring with*

$$R = CRT(\Psi_1(R), \Psi_2(R), \ldots, \Psi_s(R)) = (R_1, R_2, \ldots, R_s).$$

Let $\mathbf{v}_1, \mathbf{v}_2, \ldots, \mathbf{v}_s$ be independent non-zero vectors in R^n. Then these vectors are modular independent.

Proof If $\sum \alpha_j \mathbf{v}_j = \mathbf{0}$, then $\alpha_j \mathbf{v}_j = \mathbf{0}$ for all j. Let $\mathfrak{m}$ be the maximal ideal of R. If $\alpha_j \notin \mathfrak{m}$ for some j, then α_j is a unit. This implies that $\mathbf{v}_j = \mathbf{0}$. $\square$

We are now in a position to give the definition that we use to replace the notion of linear independence for codes over rings.

Definition 2.11 Let C be a code over a finite commutative Frobenius ring R. The codewords $\mathbf{v}_1, \mathbf{v}_2, \cdots, \mathbf{v}_k$ are called a basis of C if they are independent, modular independent, and generate C.

We can show, by the example given in [8], that modular independence does not imply independence nor does independence imply modular independence. Let $(11, 7)$ and $(3, 9)$ be vectors over $\mathbb{Z}_{12}$. Then $(11, 7)$ and $(3, 9)$ map to $(3, 3)$ and $(3, 1)$ over $\mathbb{Z}_4$ which are modular independent. Hence the vectors $(11, 7)$ and $(3, 9)$ are modular independent. But $6(11, 7) + 2(3, 9) = (0, 0)$, and $6(11, 7) = (6, 6) = 2(3, 9) \neq (0, 0)$. Therefore they are not independent. It is easy to see that $(4, 0)$ and $(0, 3)$ are independent vectors over $\mathbb{Z}_{12}$. However, $(4, 0)$ and $(0, 3)$ map to $(0, 0)$ and $(0, 3)$ over $\mathbb{Z}_4$ and map to $(1, 0)$ and $(0, 0)$ over $\mathbb{Z}_3$. Therefore, they are not modular independent.

Returning to the example which began this section, we consider the vectors $(2, 0)$ and $(0, 3)$ over $\mathbb{Z}_6$. These vectors are independent, but they are not modular independent. Hence they do not form a basis. However, the vector $(2, 3)$ is both modular independent and independent over $\mathbb{Z}_6$. Hence this single vector is the basis for this code of length 2.

We shall now show a specific case for generating free Maximum Distance Separable codes over chain rings. These ideas can be found originally in [6] and then later in more generality in [7]. Let R be a finite chain ring with the maximal ideal $\mathfrak{m} = R\gamma$ whose nilpotency index is e. This gives that $|R/\mathfrak{m}| = q = p^s$, where p is a prime and s is a positive integer. Let $M = \{\mathbf{w} \in R^r \mid | \langle \mathbf{w} \rangle | < |R| \}$. That is, M consists of vectors in R^r that have no coordinate with a unit in it. Since we are in a chain ring, we have that each coordinate of $\mathbf{w} \in M$ is a multiple of γ. This gives that

$$M = R^r. \tag{2.11}$$

We take the standard definition of linear independence. Namely, the vectors $\mathbf{v}_1, \cdots, \mathbf{v}_n \in R^r$ are linearly independent if $\sum a_i \mathbf{v}_i = \mathbf{0}$ implies $a_i = 0$ for all i.

Lemma 2.7 *Suppose that* $\mathbf{v}_1, \cdots, \mathbf{v}_{t-1} \in R^r$ *are linearly independent. If* $\mathbf{v}_t \notin \langle \mathbf{v}_1, \cdots, \mathbf{v}_{t-1}, M \rangle$, *then* $\mathbf{v}_1, \cdots, \mathbf{v}_{t-1}, \mathbf{v}_t$ *are linearly independent.*

Proof Assume $\sum_{i=1}^{t} \alpha_i \mathbf{v}_i = \mathbf{0}$. If $\alpha_t = 0$, then since $\mathbf{v}_1, \cdots \mathbf{v}_{t-1} \in R^r$ are linearly independent, we have that $\alpha_i = 0$ for all i and we have the desired result.

Next assume that α_t is a unit. This gives that $\mathbf{v}_t \in \langle \mathbf{v}_1, \mathbf{v}_2, \ldots, \mathbf{v}_{t-1} \rangle$ which is a contradiction.

Finally, assume that $\alpha_t \neq 0$ and that α_t is not a unit. Then $\alpha_t = \beta \gamma^j$ for some unit β and positive integer j. Then we have $-\beta \gamma^j \mathbf{v}_t = \sum_{i=1}^{t-1} \alpha_i \mathbf{v}_i$. Multiply both sides by γ^{e-j}. Then we have $\mathbf{0} = \sum_{i=1}^{t-1} \gamma^{e-j} \alpha_i \mathbf{v}_i$. We know that $\mathbf{v}_1, \cdots, \mathbf{v}_{t-1}$ are linearly independent, which gives that $\gamma^{e-j} \alpha_i = 0$ for all i. This implies that $\alpha_i \in \langle \gamma^j \rangle$, which gives that $\sum_{i=1}^{t} \alpha_i \mathbf{v}_i \in M$. This contradicts the assumption. □

Lemma 2.8 *Let R be a finite commutative chain ring with $|R/\mathfrak{m}| = q = p^s$, where* $\mathfrak{m} = \langle \gamma \rangle$ *is the maximal ideal, p is a prime, and s is a positive integer. Let* $M = \{ \mathbf{w} \in R^r \mid |\langle \mathbf{w} \rangle| < |R| \}$. *If $\mathbf{v}_1, \cdots, \mathbf{v}_t \in R^r$ are linearly independent, then* $|\langle \mathbf{v}_1, \cdots, \mathbf{v}_t, M \rangle| = q^t (|R|/q)^r$.

Proof We have that

$$\langle \mathbf{v}_1, \cdots, \mathbf{v}_t, M \rangle = \{ \alpha_1 \mathbf{v}_1 + \cdots + \alpha_t \mathbf{v}_t + \gamma \mathbf{w} \mid \alpha_i \in R, \mathbf{w} \in R^r \}.$$

Assume

$$\alpha_1 \mathbf{v}_1 + \cdots + \alpha_t \mathbf{v}_t + \gamma \mathbf{w}_1 = \beta_1 \mathbf{v}_1 + \cdots + \beta_t \mathbf{v}_t + \gamma \mathbf{w}_2$$

for some $\mathbf{w}_1, \mathbf{w}_2 \in R^r$. Then

$$(\alpha_1 - \beta_1) \mathbf{v}_1 + \cdots + (\alpha_t - \beta_t) \mathbf{v}_t + \gamma(\mathbf{w}_1 - \mathbf{w}_2) = \mathbf{0}.$$

Multiplying both sides by γ^{e-1}, we have

$$\gamma^{e-1} (\alpha_1 - \beta_1) \mathbf{v}_1 + \cdots + \gamma^{e-1} (\alpha_t - \beta_t) \mathbf{v}_t = \mathbf{0}.$$

Since $\mathbf{v}_1, \cdots, \mathbf{v}_t$ are linearly independent, $\alpha_i - \beta_i \in \mathfrak{m}$, which gives that $\beta_i = \alpha_i + \gamma \delta_i$ for some $\delta_i \in R$ for $i = 1, \cdots, t$. Therefore, it suffices to choose representatives of $R/R\gamma$ as coefficients of the $\mathbf{v}_i$ which counts q^t elements. Then, since $|M| = |\gamma R^r| = |R\gamma|^r = (|R|/q)^r$, the lemma follows. □

We can now imitate the Gilbert-Varshamov construction found in [13, p. 33] to obtain the following theorem.

Theorem 2.16 *Let R be a finite commutative chain ring with $|R/\mathfrak{m}| = q = p^s$, where $\mathfrak{m} = \langle \gamma \rangle$ is the maximal ideal, p is a prime, and s is a positive integer. Suppose* $\binom{n-1}{d-2} < \frac{q^{n-k}-1}{q^{d-2}-1}$. *Then there exists a free code over R of length n and rank k with minimum distance d.*

Proof Let $M = \{\mathbf{w} \in R^r \mid |\langle \mathbf{w} \rangle| < |R|\}$. To prove the theorem, we construct an $(n - k)$ by n parity check matrix H with the property that no $d - 1$ columns are linearly dependent. Set $r = n - k$. The first column can be any $\mathbf{v}_1 \in R^r$, but not in M. Suppose that we have chosen $t - 1$ columns $\mathbf{v}_1, \cdots, \mathbf{v}_{t-1} \in R^r$ so that no $d - 1$ columns are linearly dependent. Suppose there is a column $\mathbf{v}_t \notin \bigcup \langle \mathbf{v}_{i_1}, \cdots, \mathbf{v}_{i_{d-2}}, M \rangle$, where the union is taken over all possible choices of $d - 2$ columns from the $t - 1$ columns. Then no $d - 1$ columns from the t columns $\mathbf{v}_1, \cdots, \mathbf{v}_t$ are linearly dependent. Such a vector would exist if $|\bigcup \langle \mathbf{v}_{i_1}, \cdots, \mathbf{v}_{i_{d-2}}, M \rangle| < |R|^r$. Then for all $t \leq n$, we have:

$$|\bigcup \langle \mathbf{v}_{i_1}, \cdots, \mathbf{v}_{i_{d-2}}, M \rangle| \leq \binom{t-1}{d-2} |\langle \mathbf{v}_1, \cdots, \mathbf{v}_{d-2}, M \rangle| - \left(\binom{t-1}{d-2} - 1 \right) |M|$$
$$\leq \binom{n-1}{d-2} \{q^{d-2}(|R|/q)^r - (|R|/q)^r\} + (|R|/q)^r$$
$$= (|R|/q)^r \left(\binom{n-1}{d-2}(q^{d-2} - 1) + 1 \right)$$
$$< (|R|/q)^r (q^{n-k})$$
$$= |R|^r.$$

This gives the result. $\square$

This leads to the following corollary.

Corollary 2.3 *Let R be a finite chain ring with the maximal ideal $\mathfrak{m} = R\gamma$. If $q = |R/\mathfrak{m}| > \binom{n-1}{n-k-1}$ with $n - k - 1 > 0$, then there exists a Maximum Distance Separable code over R of length n containing q^{n-k+1} elements with minimum distance $n - k + 1$.*

Proof If $d = n - k + 1$, then the inequality of Theorem 2.16 becomes $\binom{n-1}{n-k-1} < \frac{q^{n-k}-1}{q^{n-k-1}-1}$. Since $q < \frac{q^{n-k}-1}{q^{n-k-1}-1} \leq q + 1$ for any n and k such that $n > k + 1$. This gives the desired result. $\square$

References

1. Atiyah, M.F., Macdonald, I.G. Introduction to Commutative Algebra. Addison-Wesley Publishing Co., Reading, Mass.-London-Don Mills, Ont. (1969)
2. Blake, I.F.: Codes over certain rings. Inf. Control **20**, 396–404 (1972)
3. Blake, I.F.: Codes over integer residue rings. Inf. Control **29**(4), 295–300 (1975)
4. Cengellenmis, Y., Dertli, A., Dougherty, S.T.: Codes over an infinite family of rings with a Gray map. Des. Codes Cryptogr. **72**(3), 559–580 (2014)
5. Dougherty, S.T.: Foundations of algebraic coding theory. Contemp. Math. **634**, 101–136 (2015)
6. Dougherty, S.T., Gulliver, T.A., Park, Y.H., Wong, J.N.C.: Optimal linear codes over $\mathbb{Z}_m$. J. Korean Math. Soc. **44**(5), 1139–1162 (2007)
7. Dougherty, S.T., Kim, J.L., Kulosman, H.: MDS codes over finite principal ideal rings. Des. Codes Cryptogr. **50**(1), 77–92 (2009)

8. Dougherty, S.T., Liu, H.: Independence of vectors in codes over rings. Des. Codes Cryptogr. **51**(1), 55–68 (2009)
9. Dougherty, S.T., Yildiz, B., Karadeniz, S,: Codes over R_k, Gray maps and their binary images. Finite Fields Appl. **17**(3), 205–219 (2011)
10. Greferath, M., Schmidt, S.E.: Finite-ring combinatorics and MacWilliams' equivalence theorem. J. Combin. Theory Ser. A **92**(1), 17–28 (2000)
11. Lam, T.Y.: A First Course in Noncommutative Rings. Graduate Texts in Mathematics, vol. 131. Springer-Verlag, New York (1991)
12. Lam, T.Y.: Lectures on Modules and Rings. Graduate Texts in Mathematics, vol. 189. Springer-Verlag, New York (1999)
13. MacWilliams, F.J., Sloane, N.J.A.: The Theory of Error-Correcting Codes. Amsterdam, North-Holland, The Netherlands (1977)
14. Matsumura, H.: Commutative Ring Theory, 2nd edn. Cambridge Studies in Advanced Mathematics, vol. 8. Cambridge University Press, Cambridge (1989). (Translated from the Japanese by M. Reid.)
15. McDonald, B.R.: Finite Rings with Identity. Marcel Dekker Inc, New York (1974)
16. Park, Y.H.: Modular independence and generator matrices for codes over $\mathbb{Z}_m$. Des. Codes Cryptogr. **50**(2), 147–162 (2009)
17. Wood, J.: Duality for modules over finite rings and applications to coding theory. Am. J. Math. **121**(3), 555–575 (1999)
18. Wood, J.: Foundations of linear codes defined over finite modules: the extension theorem and the MacWilliams identities. Lectures for the CIMPA-UNESCO-TUBITAK Summer school

Chapter 3
MacWilliams Relations

In this chapter, we prove the MacWilliams relations for codes over finite Frobenius commutative rings. These relations are one of the foundational results of algebraic coding theory.

3.1 Introduction to the MacWilliams Relations

The MacWilliams relations are one of the most important foundations of algebraic coding theory. They were first proven by F.J. MacWilliams for codes over fields in [4, 5]. These relations are able to give the weight enumerator of the orthogonal of a code from the weight enumerator of a linear code. They have numerous applications in coding theory and in the connections of coding theory to other branches of mathematics. For example, self-dual codes are codes that are equal to their orthogonals. As such, their weight enumerators are held invariant by the action of the MacWilliams relations. This leads to the natural application of invariant theory to the study of self-dual codes. See Chap. 19 of [6] for an early discussion of this application. Numerous powerful results arose from this connection. See [8] for a detailed description of the connection between self-dual codes and invariant theory.

The MacWilliams relations are so fundamental to the study of codes that it is our opinion that an alphabet is an acceptable alphabet for algebraic coding theory if and only if the alphabet admits MacWilliams relations. In [9], it is shown that the class of Frobenius rings is the class of finite commutative rings that admit such relations and this is precisely why we restrict ourselves to this class of rings. It is also possible to take finite commutative groups as alphabets since there are also MacWilliams relations for these alphabets. There are other possible alphabets, as well, but in this text, we shall restrict ourselves to Frobenius rings and commutative groups.

© The Author(s) 2017
S.T. Dougherty, *Algebraic Coding Theory Over Finite Commutative Rings*,
SpringerBriefs in Mathematics, DOI 10.1007/978-3-319-59806-2_3

We begin with the standard definition of the complete weight enumerator. We shall determine MacWilliams relations for this weight enumerator and then use this to obtain MacWilliams relations for other weight enumerators.

Definition 3.1 Let C be a code over an alphabet $A = \{a_0, a_1, \ldots, a_{r-1}\}$. The complete weight enumerator for the code C is defined as:

$$cwe_C(x_{a_0}, x_{a_1}, \ldots, x_{a_{r-1}}) = \sum_{\mathbf{c} \in C} \prod_{i=0}^{r-1} x_{a_i}^{n_i(\mathbf{c})}, \tag{3.1}$$

where there are $n_i(\mathbf{c})$ occurrences of a_i in the vector $\mathbf{c}$. The symmetrized weight enumerator of a code C over a group G is given by

$$swe_C(y_0, y_1, \ldots, y_r) = \sum_{c \in C} swt(c), \tag{3.2}$$

where $swt(c) = \prod_{i=0}^{r} x_i^{\beta_i}$ and the elements α_i and $(\alpha_i)^{-1}$ appear β_i times in the vector c. The Hamming weight enumerator is given by

$$W_C(x, y) = \sum_{c \in C} x^{n - wt_H(\mathbf{c})} y^{wt_H(\mathbf{c})} = cwe_C(x, y, y, \ldots, y). \tag{3.3}$$

For codes over the finite field of order 2, these three weight enumerators coincide. It was in this form that the MacWilliams relations first appeared in [4, 5]. For the Hamming weight enumerator, x is often set to 1, and the weight enumerator is described in terms of y.

Example 3.1 Consider the perfect code given in Example 1.3. This code is a binary code and has weight enumerator

$$W_C(x, y) = x^7 + 7x^4 y^3 + 7x^3 y^4 + y^7.$$

3.2   MacWilliams Relations for Codes Over Groups

In this section, we begin in a slightly different setting. Namely, we temporarily leave the world of codes over rings and move into codes over finite commutative groups. The reason is that the fundamental structure needed for the MacWilliams relations is the underlying additive group. Moreover, in some instances, it is useful to study additive codes. That is we want to study those codes that are simply subgroups of the underlying group structure, rather than codes that are submodules of R^n. For example, additive codes over $\mathbb{F}_4$ have received a great deal of attention because of their connection to quantum coding.

We recall that a character of G is a homomorphism $\chi : G \to \mathbb{C}^*$. Let G be a finite abelian group and fix a duality of G, that is we fix a character table of G. We have a bijective correspondence between the elements of G and those of $\widehat{G} = \{\pi | \pi$ a character of $G\}$. We note that G and $\widehat{G}$ are isomorphic as groups. However, this isomorphism is not canonical. In general, we simply choose an isomorphism and for each $\alpha \in G$, we denote the corresponding character in $\widehat{G}$ by χ_α. Note that this implies that there would be a different correspondence for a different isomorphism.

In this setting, we say that a code C over G is a subset of G^n. For a code to be linear, we require only that C be an additive subset of G^n (note that we are referring to the operation of G as an additive operation). As an example, consider the code $C = \{(0,0),(1,0),(0,1),(1,1)\} \subseteq \mathbb{F}_4^2$. This code is a subgroup of the additive group of $\mathbb{F}_4^2$, but it is not a vector space since, for example, the vector $\omega(1,1) = (\omega,\omega)$ is not in the code. Therefore, the code is not a linear code in the sense of a code over a ring.

The standard definition of the Euclidean and Hermitian inner-products do not apply here because we have only one operation. Rather, we introduce a different inner-product which will coincide with the traditional inner-products in the necessary cases.

Definition 3.2 For a code C over G, with a given isomorphism between G and $\widehat{G}$, define the orthogonal of C to be

$$C^\perp = \{(g_1, g_2, \ldots, g_n) | \prod_{i=1}^{i=n} \chi_{g_i}(c_i) = 1, \forall (c_1, \ldots, c_n) \in C\}.$$

It is imperative to understand that this orthogonal is defined with respect to a specific duality for the group. If we change the duality then we change the orthogonal for the code. In fact, a code can be equal to its dual in one duality and not in another. Despite this very general definition of the orthogonal for codes over groups, for which we shall prove MacWilliams relations, it will turn out that this description leads to MacWilliams relations for codes over Frobenius rings in a canonical way.

To each element of $\widehat{G}^n$, we associate an element of G^n with the natural correspondence. Since $(\widehat{G})^n = \widehat{G^n}$, the code $C^\perp$ is associated with the set $\{\chi \in \widehat{G^n} | \chi(\mathbf{c}) = 1$ for all $\mathbf{c} \in C\}$. This gives that $|C^\perp| = \frac{|\widehat{G}|^n}{|C|} = \frac{|G|^n}{|C|}$ and that $C = (C^\perp)^\perp$.

For a function $f : G \to A$, where A is a complex algebra, the Fourier Transform $\widehat{f}$ of f is a function $\widehat{f} : \widehat{G} \to A$ defined by

$$\widehat{f}(\pi) = \sum_{x \in G} \pi(x) f(x). \tag{3.4}$$

The following example shows how the orthogonality relation can change when the isomorphism between the group and its character group changes. When this is done the orthogonality relation can change significantly.

Example 3.2 Consider the character tables given in Example 2.6.

χ_1	0	1	ω	$1+\omega$		χ_2	0	1	ω	$1+\omega$
0	1	1	1	1		0	1	1	1	1
1	1	-1	-1	1		1	1	1	-1	-1
ω	1	-1	1	-1		ω	1	-1	-1	1
$1+\omega$	1	1	-1	-1		$1+\omega$	1	-1	1	-1

Note that each row gives the character associated to the element that indexes that row. Hence there are four characters represented in each table corresponding to the four elements of the group. For the duality generated by χ_1, we have that ω is a self-orthogonal element. Then the code $C_1 = \{0, \omega\}$ is a linear code over the additive group of $\mathbb{F}_4$ and satisfies $C_1 = C_1^{\perp_1}$ with respect to this duality. Note that this code is not linear over $\mathbb{F}_4$ as a field, nor would ω be a self-orthogonal element over the field. For the duality given by χ_2, we have that the codes $\{0, 1\}, \{0, \omega\}, \{0, 1 + \omega\}$ are all linear codes over the additive group and satisfy $C = C^{\perp_2}$. Note that in all of these cases, the usual MacWilliams relations do not apply since the codes are not linear over the field $\mathbb{F}_4$. For the duality given by χ_2, the dual code $C_1^{\perp_2} = \{0, 1+\omega\}$. In this example, the complete weight enumerator of their duals is different. We shall see that if the codes are in fact linear over the ring, the MacWilliams relations will give that the weight enumerators of both orthogonals, in this case, would have to be equal (even if the orthogonals themselves were not equal).

To find MacWilliams relations for these codes, we will need the following two well known lemmas.

Let H be a subgroup of G and let $(\widehat{G} : H) = \{\pi \in \widehat{G} \mid \pi|_H = 1\}$.

Lemma 3.1 (Poisson summation formula) *Let G be a finite group and H a subgroup of G. Let f be a function from G to a complex algebra. For every $a \in G$,*

$$\sum_{x \in H} f(a + x) = \frac{1}{|(\widehat{G} : H)|} \sum_{\pi \in (\widehat{G}:H)} \pi(-a)\widehat{f}(\pi). \qquad (3.5)$$

Lemma 3.2 *Suppose $f_i : G \rightarrow A$ are functions, $i = 1, 2, \ldots, n$, and A a complex algebra. Let $f : G^n \rightarrow A$ be given by*

$$f(x_1, \ldots, x_n) = \prod_{i=1}^{n} f_i(x_i). \qquad (3.6)$$

Then $\widehat{f} = \prod \widehat{f_i}$; i.e. if $\pi = (\pi_1, \ldots, \pi_n)$ in $\widehat{G^n} = \prod_{i=1}^{n} \widehat{G}$, then $\widehat{f}(\pi) = \prod_{i=1}^{n} \widehat{f_i}(\pi_i)$.

Let $f_i(c_i) = x_{c_i}$ and $f(x) = \prod_{i=1}^{n} f_i$. Then apply the previous two lemmas, which gives that for a subgroup H of G,

$$\sum_{x \in H} f(x) = \frac{1}{|(\widehat{G} : H)|} \sum_{\pi \in (\widehat{G}:H)} \widehat{f}(\pi). \qquad (3.7)$$

Then noting that $\widehat{f}(\pi) = \sum_{x \in G} \pi(x) f(x)$ gives that the action of the matrix T on the weight enumerator gives us the MacWilliams relations, where T is indexed by the elements of the group and is defined as $T_{\alpha_i, \alpha_j} = \chi_{\alpha_i}(\alpha_j)$. For a vector $\mathbf{v}$ we let $T \cdot \mathbf{v} = (T\mathbf{v}^t)^t$.

For an element $a \in G$, let $[a]$ denote the equivalence class formed under the relation where $a \equiv a'$ if and only if $a = a'$ of $a^{-1} = (a')$, where a^{-1} is the inverse with respect to the operation of the group. Construct the matrix S indexed by $(G/\equiv)$ where $S_{[a],[b]} = T_{a,b} + T_{a,-b}$. Let s' be the number of equivalence classes. Now we can state the MacWilliams relations for groups.

Theorem 3.1 *Let C be a code over G and let $|G| = s$, with weight enumerator $cwe_C(x_0, x_1, \ldots, x_{s-1})$. Then, the complete weight enumerator of the orthogonal is given by*

$$cwe_{C^\perp} = \frac{1}{|C|} cwe_C(T \cdot (x_0, x_1, \ldots, x_{s-1})), \qquad (3.8)$$

$$swe_{C^\perp} = \frac{1}{|C|} swe_C(S \cdot (x_0, x_1, \ldots, x_{s'-1})), \qquad (3.9)$$

and

$$W_{C^\perp} = \frac{1}{|C|} W_C(x + (s - 1)y, x - y). \qquad (3.10)$$

Proof The first equation follows from the discussion above. The second equation follows easily from specializing the variables. To get the Hamming weight enumerator, notice that specializing the variables gives

$$\sum_{\alpha \in G} \chi_\alpha(\beta) x_\beta = x + (\sum_{\alpha \neq 0} \chi_\alpha(\beta)) y, \qquad (3.11)$$

where 0 is the identity of the group. If $\beta = 0$, then $\sum_{\alpha \neq 0} \chi_\alpha(\beta) = s - 1$. If $\beta \neq 0$ then $\sum_{\alpha \neq 0} \chi_\alpha(\beta) = -1$. $\square$

We shall now generalize the MacWilliams relations to the g-fold joint weight for codes over Frobenius rings.

Definition 3.3 Let G be a finite commutative group and let $C_1, \ldots, C_g$ be additive codes over G. The complete joint weight enumerator of genus g for codes $C_1, \ldots, C_g$ of length n is defined as

$$\Im_{C_1,\ldots,C_g}(X_{\mathbf{a}} : \mathbf{a} \in G^g) = \sum_{(\mathbf{c}_1,\ldots,\mathbf{c}_g) \in C_1 \times \cdots \times C_g} \prod_{\mathbf{a} \in G^g} X_{\mathbf{a}}^{n_{\mathbf{a}}(\mathbf{c}_1,\ldots,\mathbf{c}_g)},$$

where

$$\mathbf{c}_l = (c_{l1}, \cdots, c_{ln}), 1 \le l \le g$$

and

$$n_{\mathbf{a}}(\mathbf{c}_1, \cdots, \mathbf{c}_g) = |\{m \mid (c_{1m}, \cdots, c_{gm}) = \mathbf{a}, 1 \le m \le n\}|.$$

Fix a duality T for the group G. The proof of the following is a straightforward computation similar to the proof for the usual MacWilliams relations.

Corollary 3.1 *Let $C_1, \cdots, C_g$ be additve codes over a finite group G and let $\tilde{C}_l$ denote either C_l or $C_l^{\perp}$. Then*

$$\Im_{\tilde{C}_1,\cdots,\tilde{C}_g}(X_{\mathbf{a}}) = \frac{1}{\prod_{l=1}^{g} |C_l|^{\delta_{\tilde{C}_l}}} \cdot (\otimes_{l=1}^{g} T^{\delta_{\tilde{C}_l}})\Im_{C_1,\cdots,C_g}(X_{\mathbf{a}}), \qquad (3.12)$$

where

$$\delta_{\tilde{C}_l} = \begin{cases} 0 & \text{if } \tilde{C}_l = C_l, \\ 1 & \text{if } \tilde{C}_l = C_l^{\perp}. \end{cases}$$

3.3   MacWilliams Relations for Codes Over Rings

We can now use the results for codes over groups to produce MacWilliams relations for codes over Frobenius rings.

Lemma 3.3 *Let R be a finite commutative Frobenius ring with $\widehat{R} = \langle \chi \rangle$. Define the following function $F : R^n \to \widehat{R^n}$ by*

$$F(\mathbf{v}) = \chi_{\mathbf{v}}, \text{ where } \chi_{\mathbf{v}}(\mathbf{w}) = \chi([\mathbf{v}, \mathbf{w}]). \qquad (3.13)$$

Proof It is clear that the map is a homomorphism. We have that

$$ker(F) = \{\mathbf{v} \mid \chi([\mathbf{v}, \mathbf{w}]) = 1 \text{ for all } \mathbf{w} \in R^n\}.$$

Since $e_i = (0, 0, \ldots, 0, 1, 0, \ldots, 0) \in R^n$, we have that $ker(F)$ is trivial and therefore F is an injection. Moreover, $|R^n| = |\widehat{R^n}|$, which gives that the map is a bijection and hence an isomorphism. $\qquad \square$

Note that we are heavily using the fact that the ring is Frobenius in the definition of this map since otherwise we would not have a generating character χ to define it in this manner.

Let C be a linear code in R^n. Let $C^\perp$ be the standard orthogonal for a code over a ring. Let $\mathcal{L}(C)$ be the orthogonal for C as a subgroup of the additive group of R^n with the duality given by the character χ, namely $\chi_a(b) = \chi(ab)$. We have from Lemma 3.3 that $F(C^\perp) = \mathcal{L}(C)$ which together with the group theoretic MacWilliams relations given in Theorem 3.1 gives the following MacWilliams relations for codes over finite commutative Frobenius rings.

Theorem 3.2 *Let C be a linear code over a finite commutative Frobenius ring R. Define $T_{a,b} = \chi(ab)$, where χ is the generating character associated with R. Let S be the matrix indexed by the equivalence classes formed by the relation where $a \equiv a'$ if and only if $a = \pm a'$, and $S_{[a],[b]} = T_{a,b} + T_{a,b'}$. Then we have the following:*

$$cwe_{C^\perp} = \frac{1}{|C|} cwe_C(T \cdot (x_0, x_1, \ldots, x_{s-1})), \tag{3.14}$$

$$swe_{C^\perp} = \frac{1}{|C|} swe_C(S \cdot (x_0, x_1, \ldots, x_{s'-1})). \tag{3.15}$$

Note that we are not saying that there is a unique way to express the MacWilliams relations since it depends on the generating character which is not unique for a given ring. However, different matrices will still give the same weight enumerator for the orthogonal.

The following was first proven by F.J. MacWilliams in [4, 5]. There it was proven for codes over finite fields. Here we can extend the proof to codes over finite commutative Frobenius rings.

Theorem 3.3 *Let R be a finite commutative Frobenius ring with $|R| = r$. Let C be a linear code over R. Then*

$$W_{C^\perp}(x, y) = \frac{1}{|C|} W_C(x + (r - 1)y, x - y). \tag{3.16}$$

Proof The result follows from Theorem 3.2 by taking the matrix T given in that theorem and by adding all non-zero columns. In the first row, adding all non-zero columns gives $(r - 1)$ since every element is 1. Then in any other row (since all non-zero elements have the same Hamming weight), we get -1 when summing the columns since $\sum_{b \in R} \chi(ab) = 0$ for all non-zero $a \in R$ and $\chi(a0) = 1$.

Hence, the matrix that gives the MacWilliams relations is:

$$\begin{pmatrix} 1 & (r-1) \\ 1 & -1 \end{pmatrix}, \tag{3.17}$$

and this gives the result. $\square$

It is unclear where the next corollary first appeared. It is implicit in [9] but does not appear there. However, it is one of the most important consequences of the MacWilliams relations.

Corollary 3.2 *If C is a linear code over a finite commutative Frobenius ring R, with $|R| = r$, then $|C||C^\perp| = |R^n|$.*

Proof Consider Eq. 3.16 and set $x = 1$ and $y = 1$. Then we have

$$|C^\perp| = \frac{1}{|C|} r^n \tag{3.18}$$

which gives $|C||C^\perp| = |R^n|$. $\square$

One of the main uses of this corollary is that if we have a self-orthogonal code C with $|C| = \sqrt{|R^n|}$, then C is self-dual. This corollary can also be used as a tool to show a ring is not Frobenius. Namely, if a ring has an ideal $\mathfrak{a}$ where its orthogonal does not have cardinality $\frac{|R|}{|\mathfrak{a}|}$, then the ring is not Frobenius. We shall show an example where this fails when the ring is not Frobenius.

Example 3.3 Let $R = \mathbb{F}_2[x, y]/(x^2, y^2, xy)$. We can write the elements of R as $R = \{0, 1, x, y, 1 + x, 1 + y, x + y, 1 + x + y\}$.

The maximal ideal is $\mathfrak{m} = \{0, x, y, x + y\}$. Hence, this ideal is a code of length 1. Its orthogonal is $\mathfrak{m}^\perp = \mathfrak{m} = \{0, x, y, x + y\}$. This gives that $\mathfrak{m}$ is a self-dual code of length 1. However $|\mathfrak{m}||\mathfrak{m}^\perp| = 16 \neq |R| = 8$. This implies that there cannot be MacWilliams relations for this ring, since if there were then $|C||C^\perp|$ would have to be $|R|^n$.

The MacWilliams relations are an extremely powerful tool. We shall exhibit one of their classical applications.

Example 3.4 Consider the Hamming codes $H(2, r)$ given in Example 1.7. The orthogonal to this code has dimension r and length $2^r - 1$. Since every possible non-zero column is represented in the generator matrix, then the sum of any subset of rows produces a vector with weight 2^{r-1}. This gives that the weight enumerator of $H(2, r)^\perp$ is

$$W_{H(2,r)^\perp} = x^{2^r - 1} + x^{2^{r-1}-1} y^{2^{r-1}}.$$

Then applying the MacWilliams relations gives the weight enumerator of the binary Hamming codes. Namely,

$$W_{H(2,r)} = (x + y)^{2^r - 1} + (x + y)^{2^{r-1}-1} (x - y)^{2^{r-1}}.$$

Corollary 3.3 *Let T be the matrix that gives the MacWilliams relations for a finite commutative Frobenius ring R with $|R| = r$. Then $T^2 = rM$ where M is a monomial matrix corresponding to a permutation of the elements of R.*

Proof For any linear code C we have that $(C^\perp)^\perp = C$. This gives that applying the MacWilliams relations twice will result in the weight enumerator of the original code. This implies that $(\frac{1}{\sqrt{r}} T)(\frac{1}{\sqrt{r}} T)$ must be a monomial matrix. The result follows.

Example 3.5 For $\mathbb{Z}_4$, the matrix T which gives the MacWilliams relations is:

$$\begin{pmatrix} 1 & 1 & 1 & 1 \\ 1 & i & -1 & -i \\ 1 & -1 & 1 & -1 \\ 1 & -i & -1 & i \end{pmatrix}. \tag{3.19}$$

Then we have

$$T^2 = 4 \begin{pmatrix} 1 & 0 & 0 & 0 \\ 0 & 0 & 0 & 1 \\ 0 & 0 & 1 & 0 \\ 0 & 1 & 0 & 0 \end{pmatrix}. \tag{3.20}$$

MacWilliams relations for non-Hamming weight enumerators can also be found. See [2] for the MacWilliams relations for the Rosenbloom-Tsfasman metric.

3.4 A Practical Guide to the MacWilliams Relations

We shall now show how to construct MacWilliams relations for specific rings. As usual with commutative rings and coding theory, one of the most powerful tools is the application of the Chinese Remainder Theorem.

Theorem 3.4 *Let R be a finite commutative Frobenius ring with $R = CRT(R_1, R_2, \ldots, R_s)$, where each R_i is a local ring. Let χ_{R_i} be the generating character for R_i. Then the character χ for R defined by*

$$\chi(a) = \prod \chi_{R_i}(a_i), \tag{3.21}$$

where $a = CRT(a_1, a_2, \ldots, a_s)$, is a generating character for R.

Proof If χ were not a generating character, then by Theorem 2.4, it would be trivial on an ideal of R. Then there would be an i such that χ_{R_i} is trivial on an ideal of R_i, contradicting that χ_{R_i} is a generating character. Hence, χ is a generating character. $\square$

This theorem allows us to focus on local rings since we know that any finite commutative ring is isomorphic via the Chinese Remainder Theorem to a product of local rings. With this in mind, let R be a finite local commutative Frobenius ring with maximal ideal $\mathfrak{m}$. Then it follows that $\mathfrak{m}^{\perp}$ is the unique minimal ideal of R where $\mathfrak{m}$ is the unique maximal ideal.

Lemma 3.4 *Let R be a finite local commutative Frobenius ring with maximal ideal $\mathfrak{m}$. If χ is a character of R that is not trivial on $\mathfrak{m}^{\perp}$, then χ is a generating character for $\widehat{R}$.*

Proof We know that $\mathfrak{m}^\perp$ is the unique minimal ideal. This means that $\mathfrak{m}^\perp$ is contained in every non-trivial ideal of R. Hence if χ is non-trivial on $\mathfrak{m}^\perp$, then it is non-trivial on every ideal of R. This gives that it is a generating character. $\qquad\square$

Lemma 3.4 gives an easy way to find a generating character for any finite local commutative Frobenius ring. Namely, we simply find a character that is not trivial on the unique minimal ideal.

This lemma tells us a lot more about the MacWilliams relations for codes over rings. Namely, there is not a unique way to give the matrix T for a specific Frobenius ring. Rather the MacWilliams relations apply to any linear code over the ring but the matrix T depends on the choice of the generating character which as we see from the previous lemma is not necessarily unique. However, there is still only one matrix that applies the MacWilliams relations for the Hamming weight enumerator. That is, every possible matrix T still collapses to the same matrix for the Hamming weight enumerator.

We shall give some examples of the matrix T, which gives the MacWilliams relations for various rings.

- Consider the ring $\mathbb{Z}_n$. The classical Chinese Remainder Theorem gives that $\mathbb{Z}_n \cong \mathbb{Z}_{p_1^{e_1}} \times \mathbb{Z}_{p_2^{e_2}} \times \cdots \times \mathbb{Z}_{p_s^{e_s}}$, where the p_i are distinct primes. The ring $\mathbb{Z}_{p_i^{e_i}}$ is a local ring with maximal ideal $\langle p_i \rangle$ and minimal ideal $\langle p_i^{e_i-1} \rangle$. Let $\chi_{\mathbb{Z}_{p_i^{e_i}}}(a) = \eta_i^a$, where η_i is a primitive $p_i^{e_i}$-th root of unity. Then $\eta_i^{p_i^{e_i-1}} \neq 1$ and so $\chi_{\mathbb{Z}_{p_i^{e_i}}}$ is non-trivial on the minimal ideal and therefore is a generating character. Then $\chi_{\mathbb{Z}_n} = \prod \chi_{\mathbb{Z}_{p_i^{e_i}}}$, which is realized as $\chi_{\mathbb{Z}_n}(a) = \eta^a$ where η is a primitive n-th root of unity.
- Consider the Galois ring $\mathbb{Z}_{p^e}[x]/\langle q(x) \rangle$, where $q(x)$ is an irreducible polynomial over $\mathbb{Z}_{p^e}$ of degree k and p is a prime. Here any element is of the form $a_0 + a_1 x + \cdots + a_{k-1} x^{k-1}$. Then

$$\chi(a_0 + a_1 x + \cdots + a_{e-1} x^{e-1}) = \xi_{p^e}^{\sum a_i} \qquad (3.22)$$

is a generating character for $\widehat{\mathbb{Z}_{p^e}[x]/\langle q(x) \rangle}$ where ξ_{p^e} is a primitive p^e-th root of unity. Of course, when $e = 1$ this gives us the class of finite fields.
- Consider the rings $R_k = \mathbb{F}_2[u_1, u_2, \ldots, u_k]$, where $u_i^2 = 0$ and $u_i j_j = u_j u_i$ for all i, j. Then for $A \subseteq \{1, 2, \ldots, k\}$ we denote $u_A = \prod_{i \in A} u_i$ and each element can be written as $\sum_{A \subset \mathcal{P}(\{1,2,\ldots,k\})} \alpha_A u_A$, where $\alpha_A \in \mathbb{F}_2$. Then

$$\chi\left(\sum_{A \subseteq \mathcal{P}(\{1,2,\ldots,k\})} \alpha_A u_A \right) = -1^{\sum \alpha_A} \qquad (3.23)$$

is a generating character for $\widehat{R_k}$.

- Let R be a finite chain ring with maximal ideal $\langle \gamma \rangle$ where $R/\langle \gamma \rangle$ is isomorphic to $\mathbb{F}_q$. Let χ_q be the generating character for $\mathbb{F}_q$. We have that $\langle \gamma^{e-1} \rangle$ is the minimal ideal. Then let χ be defined by

$$\chi(a_0 + a_1\gamma + \cdots + a_{e-1}\gamma^{e-1}) = \prod \chi_q(a_i). \qquad (3.24)$$

It follows that χ is not minimal on $\langle \gamma^{e-1} \rangle$ and therefore χ is a generating character for $\widehat{R}$.

In general, we are interested in determining the generating character of local rings, since by using Theorem 3.4, we can then determine the MacWilliams relations for any finite Frobenius commutative ring. In [7], it is shown that the smallest local Frobenius ring that is not a chain ring has order 16. Hence, the previous discussion gives the MacWilliams relations for all rings of order less than 16. Additionally in [7], the local Frobenius rings of order 16 were classified. In [1], the generating character for all of these rings is given. We give them in Table 3.1. In the table, $\eta = e^{\frac{2\pi i}{8}}$ and $\zeta = e^{\frac{2\pi i}{16}}$.

Table 3.1 Generating characters for local Frobenius rings of order 16

Ring	Additive structure	Generating character
$\mathbb{F}_{16} \cong \frac{\mathbb{F}_2[x]}{\langle x^4+x+1 \rangle}$	$\mathbb{Z}_2 \times \mathbb{Z}_2 \times \mathbb{Z}_2 \times \mathbb{Z}_2$	$\chi(a+bx+cx^2+dx^3) = (-1)^{a+b+c+d}$
$\frac{\mathbb{F}_2[x]}{\langle x^4 \rangle}$	$\mathbb{Z}_2 \times \mathbb{Z}_2 \times \mathbb{Z}_2 \times \mathbb{Z}_2$	$\chi(a+bx+cx^2+dx^3) = (-1)^{a+b+c+d}$
$\frac{\mathbb{F}_4[x]}{\langle x^2 \rangle} \cong \frac{\mathbb{F}_2[u,v]}{\langle u^2+u+1,v^2 \rangle}$	$\mathbb{Z}_2 \times \mathbb{Z}_2 \times \mathbb{Z}_2 \times \mathbb{Z}_2$	$\chi(a+bu+cv+duv) = (-1)^{a+b+c+d}$
$\frac{\mathbb{F}_2[u,v]}{\langle u^2,v^2 \rangle}$	$\mathbb{Z}_2 \times \mathbb{Z}_2 \times \mathbb{Z}_2 \times \mathbb{Z}_2$	$\chi(a+bu+cv+duv) = (-1)^{a+b+c+d}$
$\frac{\mathbb{F}_2[u,v]}{\langle u^2+v^2,uv \rangle}$	$\mathbb{Z}_2 \times \mathbb{Z}_2 \times \mathbb{Z}_2 \times \mathbb{Z}_2$	$\chi(a+bu+cv+du^2) = (-1)^{a+b+c+d}$
$GR(2^2,2) \cong \frac{\mathbb{Z}_4[x]}{\langle x^2+x+1 \rangle}$	$\mathbb{Z}_4 \times \mathbb{Z}_4$	$\chi(a+bx) = i^{a+b}$
$\frac{\mathbb{Z}_4[x]}{\langle x^2-2 \rangle}$	$\mathbb{Z}_4 \times \mathbb{Z}_4$	$\chi(a+bx) = i^{a+b}$
$\frac{\mathbb{Z}_4[x]}{\langle x^2-2x-2 \rangle}$	$\mathbb{Z}_4 \times \mathbb{Z}_4$	$\chi(a+bx) = i^{a+b}$
$\frac{\mathbb{Z}_4[x]}{\langle x^2 \rangle}$	$\mathbb{Z}_4 \times \mathbb{Z}_4$	$\chi(a+bx) = i^{a+b}$
$\frac{\mathbb{Z}_4[x]}{\langle x^2-2x \rangle}$	$\mathbb{Z}_4 \times \mathbb{Z}_4$	$\chi(a+bx) = i^{a+b}$
$\frac{\mathbb{Z}_4[x]}{\langle x^3-2,2x \rangle}$	$\mathbb{Z}_4 \times \mathbb{Z}_2 \times \mathbb{Z}_2$	$\chi(a+bx+cx^2) = i^a(-1)^{b+c} = i^{a+2b+2c}$
$\frac{\mathbb{Z}_4[x,y]}{\langle x^2,xy-2,y^2,2x,2y \rangle}$	$\mathbb{Z}_4 \times \mathbb{Z}_2 \times \mathbb{Z}_2$	$\chi(a+bx+cy) = i^a(-1)^{b+c} = i^{a+2b+2c}$
$\frac{\mathbb{Z}_4[x,y]}{\langle x^2-2,xy-2,y^2,2x,2y \rangle}$	$\mathbb{Z}_4 \times \mathbb{Z}_2 \times \mathbb{Z}_2$	$\chi(a+bx+cy) = i^a(-1)^{b+c} = i^{a+2b+2c}$
$\frac{\mathbb{Z}_8[x,y]}{\langle x^2-4,2x \rangle}$	$\mathbb{Z}_8 \times \mathbb{Z}_2$	$\chi(a+bx) = \eta^a(-1)^b = \eta^{a+4b}$
$\mathbb{Z}_{16}$	$\mathbb{Z}_{16}$	$\chi(a) = \zeta^a$

References

1. Dougherty, S.T., Saltürk, E., Szabo, S.: On codes over local Frobenius rings: generator matrices, generating characters and MacWilliams Identities (to appear)
2. Dougherty, S.T., Skriganov, M.M.: MacWilliams duality and the rosenbloom-tsfasman metric. Moscow Math. J. **2**(1), 83–99 (2002)
3. Greferath, M., Schmidt, S.E.: Finite-ring combinatorics and MacWilliams' equivalence theorem. J. Combin. Theory Ser. A **92**(1), 17–28 (2000)
4. MacWilliams, F.J.: Combinatorial problems of elementary group theory, Ph.D. thesis, Harvard University (1961)
5. MacWilliams, F.J.: A theorem on the distribution of weights in a systematic code. Bell Syst. Tech. J. **42**, 79–94 (1963)
6. MacWilliams, F.J., Sloane, N.J.A.: The Theory of Error-Correcting Codes. North-Holland, Amsterdam (1977)
7. Martnez-Moro, E., Szabo, S.: On codes over local Frobenius non-chain rings of order 16, non-commutative rings and their applications. Contemp. Math. **634**, 227–241 (2015)
8. Nebe, G., Rains, E.M., Sloane, N.J.A.: Self-Dual Codes And Invariant Theory, Algorithms and Computation in Mathematics, 17. Springer, Heidelberg (2006)
9. Wood, J.: Duality for modules over finite rings and applications to coding theory. Am. J. Math. **121**(3), 555–575 (1999)

Chapter 4
Families of Rings

The study of codes over rings began in earnest with studying codes over the ring $\mathbb{Z}_4$ in [20, 21]. The central result of this work was that certain non-linear binary codes could be viewed as linear codes over $\mathbb{Z}_4$ via a non-linear Gray map ϕ. It was primarily this result which started an intense study of codes over rings, especially codes where there exists a Gray map from the ring to the Hamming space. Generally, the rings that were first studied were the integer modular rings and rings which were similar to those such as chain rings and principal ideal rings. Certain applications, like the connection to unimodular lattices, caused other rings to become of interest to coding theorists. For example, the ring $\mathbb{Z}_{2k}$ has a natural connection to real lattices, and the ring $\mathbb{F}_2 + u\mathbb{F}_2$ has a connection to complex lattices. In this chapter, we shall study families of rings that are of particular interest to coding theorists and, in many cases, an associated Gray map with these families. Throughout this work, we shall say that a map is a Gray map if it is a distance preserving map to $\mathbb{F}_2^N$ for some N. Often, the distance in the ring is the induced distance given by the Hamming distance in $\mathbb{F}_2^N$. The map is then distance preserving by definition. The primary benefit of this situation is that we are largely concerned with the code that is the image under the Gray map. In this way, if the distance defined in this manner is non-canonical, it is not a concern.

4.1 Rings of Order 4

The first rings that were studied heavily in coding theory were the commutative rings of order 4. There are four commutative rings of order 4. We shall use the names associated with them in the coding theory literature to be consistent with the existing literature. These rings are $\mathbb{Z}_4$, $\mathbb{F}_2[u]/\langle u^2 \rangle$, $\mathbb{F}_2[v]/\langle v^2 + v \rangle$, and $\mathbb{F}_4 = \mathbb{F}_2[\omega]/\langle \omega^2 + \omega + 1 \rangle$. In the literature, $\mathbb{F}_2[u]/\langle u^2 \rangle$ is usually called $\mathbb{F}_2 + u\mathbb{F}_2$ and $\mathbb{F}_2[v]/\langle v^2 + v \rangle$ is called $\mathbb{F}_2 + v\mathbb{F}_2$. Each of these rings has a Gray map associated with it, where a Gray map is a distance preserving map to the binary Hamming space. For $\mathbb{Z}_4$, we write each element as $a + b2$, with $a, b, \in \mathbb{F}_2$ and define the Gray map $\phi_4 : \mathbb{Z}_4 \to \mathbb{F}_2^2$ as

© The Author(s) 2017
S.T. Dougherty, *Algebraic Coding Theory Over Finite Commutative Rings*,
SpringerBriefs in Mathematics, DOI 10.1007/978-3-319-59806-2_4

$\phi_4(a + b2) = (b, b + a)$. Likewise for $\mathbb{F}_2 + u\mathbb{F}_2$, we write each element as $a + bu$ and define the Gray map $\phi_u : \mathbb{F}_2 + u\mathbb{F}_2 \to \mathbb{F}_2^2$ as $\phi_u(a + b2) = (b, b + a)$. These two maps are similar but there is a significant difference. Namely, for $\mathbb{Z}_4$ the map is non-linear, but for $\mathbb{F}_2 + u\mathbb{F}_2$ the map is linear. The ring $\mathbb{F}_2 + v\mathbb{F}_2$ is isomorphic via the Chinese Remainder Theorem to $\mathbb{F}_2^2$. Specifically, $a + bv = (a + b)v + a(v + 1)$ and we define the Gray map as $\phi_v(a + bv) = (a, a + b)$. This map is linear as well, as it is the inverse of the canonical isomorphism from the Chinese Remainder Theorem. Finally, for the field of order 4, the standard projection is the corresponding map. We define $\phi_\omega(a + b\omega) = (a, b)$. These maps are realized in the following table:

$\mathbb{F}_2^2$	$\mathbb{Z}_4$	$\mathbb{F}_2 + u\mathbb{F}_2$	$\mathbb{F}_2 + v\mathbb{F}_2$	$\mathbb{F}_4$
00	0	0	0	0
01	1	1	v	ω
11	2	u	1	$1 + \omega$
10	3	$1 + u$	$1 + v$	1

These maps are extended in the natural way to be a map from R^n to $\mathbb{F}_2^{2n}$. This allows us to examine binary codes which are the images of linear codes over these rings. For the ring $\mathbb{Z}_4$, we obtain binary codes which may not be linear but which do have a group structure inherited from the group structure of the code in $\mathbb{Z}_4^n$. Given the results in [5], it seems that this should have been noticed earlier. The primary result which makes the Gray map for $\mathbb{Z}_4$ so interesting is that certain non-linear binary codes can behave like linear codes with respect to the all important MacWilliams relations. Guided by this, we make the following definition.

Definition 4.1 Let R be a finite commutative Frobenius ring and let $\phi : R \to \mathbb{F}_2^s$ be a Gray map. We define the Lee weight of an element $a \in R$ as $wt_L(a) = wt_H(\phi(a))$.

For example, the elements 1 and 3 in $\mathbb{Z}_4$ have Lee weight 1, and the element 2 has Lee weight 2. This definition allows us to define the Lee weight enumerator as follows.

Definition 4.2 Let C be a code over a finite commutative Frobenius ring with an associated Gray map. Then

$$L_C(x, y) = \sum_{c \in C} x^{N - wt_L(c)} y^{wt_L(c)}, \tag{4.1}$$

where N is the length of $\phi(C)$.

Note that the Lee weight enumerator of a code C is identical to the Hamming weight enumerator of its image under the Gray map. That is, $L_C(x, y) = W_{\phi(C)}(x, y)$.

In [20, 21], the authors prove and make substantial use of the following theorem.

Theorem 4.1 *Let C be a linear code over $\mathbb{Z}_4$. Then*

$$L_{C^\perp}(x, y) = \frac{1}{|C|} L_C(x + y, x - y). \tag{4.2}$$

In other words, the Lee weight enumerator for linear codes over $\mathbb{Z}_4$ follows the same MacWilliams relations as a binary linear code even though its image may not be a linear code. The same MacWilliams relations will hold for codes over $\mathbb{F}_2 + u\mathbb{F}_2$, which we will prove later for the family of codes that generalizes this ring.

We have already defined the Hamming and Lee weights for codes over rings of order 4. There are two additional weights we shall consider. The first is the Euclidean weight which is defined on $\mathbb{Z}_4$ and $\mathbb{F}_2 + u\mathbb{F}_2$. The weights are 0, 1, 4, 1 for 0, 1, 2, 3 and $0, 1, u, 1+u$ respectively. For the ring $\mathbb{F}_2 + v\mathbb{F}_2$ the Bachoc weights are 0, 2, 1, 2 for $0, v, 1, 1+v$ respectively. The Euclidean and Bachoc weights are derived from the corresponding weight related to the corresponding weights in the lattices constructed from these codes, see [1, 2, 9]. The minimum distance of the code with respect to these weights are denoted by $d_H(C), d_L(C), d_E(C)$ and $d_B(C)$.

Recall that the rank of a linear code C, denoted by rank(C), is the minimum number of generators of C and the free rank denoted by f-rank(C) is the maximum of the ranks of R-free submodules of C. The code is said to be free when the free rank and the rank coincide. The cardinality of a linear code over a ring of order 4 is $4^{\text{f-rank}(C)} 2^{\text{rank}(C) - \text{f-rank}(C)}$.

We know that any binary code C satisfies $d_H(C) \leq n - \log_2 |C| + 1$. This gives the following theorem which first appeared in [13].

Theorem 4.2 *Let C be a possibly non-linear code of length n over a commutative ring of order 4 and minimum Lee distance d_L. Then*

$$d_L \leq 2n - \log_2 |C| + 1. \tag{4.3}$$

If we let k_1 be f-rank(C) and $k_2 = $ f-rank$(C) - $ rank(C), then for a linear code we can rewrite Eq. 4.3 as

$$\frac{d_L - 1}{2} \leq n - k_1 - \frac{k_2}{2}. \tag{4.4}$$

A code that meets this bound is known as a Maximum Lee Distance Separable (MLDS) code. It is immediate that the image of an MLDS code is a binary MDS code. It is well known that for binary codes the only MDS codes are $\langle \mathbf{j} \rangle$ with parameters $[n, 1, n]$, $\langle \mathbf{j} \rangle^\perp$ with parameters $[n, n-1, 2]$, and $\mathbb{F}_2^n$ with parameters $[n, n, 1]$, where $\mathbf{j}$ denotes the all-one vector. Therefore, the only MLDS codes are the preimages of these codes under the Gray map.

We shall now describe a general approach to Singleton type bounds for codes over rings of order 4, which first appeared in [13].

Lemma 4.1 *Let C be a linear code of length n over a commutative ring of order 4. Then*

$$\text{rank}(C) + \text{f-rank}(C^\perp) = n. \tag{4.5}$$

Proof For the finite field, the result is trivial. For $\mathbb{Z}_4$ and $\mathbb{F}_2 + u\mathbb{F}_2$, the rings are chain rings, and it follows from Theorem 3.1 in [22]. For the ring $\mathbb{F}_2 + v\mathbb{F}_2$, the result follows from the fact that the ring is isomorphic to $\mathbb{F}_2 \times \mathbb{F}_2$ via the Chinese

Remainder Theorem. Therefore, if C is a code over this ring then $C = CRT(C_1, C_2)$ and $C^{\perp} = CRT(C_1^{\perp}, C_2^{\perp})$. Then we have that $\mathrm{rank}(C) = \max\{\dim(C_1), \dim(C_2)\}$ and f-$\mathrm{rank}(C) = \min\{\dim(C_1), \dim(C_2)\}$, which gives the result. $\square$

Let R be a commutative ring of order 4 and let D be a submodule of R^n. Let $M \subseteq \{1, 2, \ldots, n\}$. Let $D(M) = \{\mathbf{x} \in D \mid \mathrm{supp}(\mathbf{x}) \subseteq M\}$ and let $D^* = \mathrm{Hom}_R(D, R)$.

It is immediate that $D(M) = D \cap R^n(M)$ is a submodule of R^n and $|R^n(M)| = 4^{|M|}$. There exists an isomorphism which gives $D^* \cong D$ and there is an R-homomorphism $g : R^n \longrightarrow D^*$ defined by

$$g(\mathbf{y}) = (\hat{\mathbf{y}} : x \mapsto [\mathbf{x}, \mathbf{y}]).$$

The map g is surjective since R is Frobenius, (see [26] for details). Since R is a commutative Frobenius ring, we have the following basic exact sequence (see [29]). Let C be a code of length n over R and $M \subseteq \{1, 2, \ldots, n\} = N$. Then there is an exact sequence of R-modules:

$$0 \longrightarrow C^{\perp}(M) \overset{\mathrm{inc}}{\longrightarrow} R^n(M) \overset{g}{\longrightarrow} C^* \overset{\mathrm{res}}{\longrightarrow} C(N - M)^* \longrightarrow 0. \qquad (4.6)$$

Here the inc denotes the inclusion map and res denotes the restriction map.

Define $n_r(\mathbf{x})$ to be $n_r(\mathbf{x}) := |\{i \mid x_i = r\}|$. Let $a_r, r \in R - \{0\}$, be positive integers, set $a_0 = 0$, and let $w(\mathbf{x}) := \sum_{r \in R} a_r n_r(\mathbf{x})$. When $a_r = 1$, this is the Hamming weight.

Let $A := \max\{a_r \mid r \in R\}$. If $R = \mathbb{Z}_4 = \{0, 1, 2, 3\}$, then setting $a_1 = a_3 = 1$ and $a_2 = 2$ yields the Lee weight, while setting $a_1 = a_3 = 1$ and $a_2 = 4$ yields the Euclidean weight, and if $R = \mathbb{F}_2 + v\mathbb{F}_2 = \{0, 1, v, 1 + v\}$, then setting $a_1 = 1$ and $a_v = a_{1+v} = 2$ yields the Bachoc weight.

Let G be a general weight and let d_G be $d_G := \min\{w(\mathbf{x}) \mid \mathbf{x} \in C - \{\mathbf{0}\}\}$. That is d_G is the minimum weight with respect to that weight.

We note that $w(\mathbf{x}) \leq A|\mathrm{supp}(\mathbf{x})|$.

This takes us to our main general bound.

Theorem 4.3 *Let R be a commutative ring of order 4 and let C be a linear code of length n over R with minimum weight d_G and maximum a_r-value A. Then*

$$\left\lfloor \frac{d_G - 1}{A} \right\rfloor \leq n - \mathrm{rank}(C). \qquad (4.7)$$

Proof In the above exact sequence, replace C with $C^{\perp}$. This gives the following exact sequence:

$$0 \longrightarrow C(M) \overset{\mathrm{inc}}{\longrightarrow} R^n(M) \overset{g}{\longrightarrow} (C^{\perp})^* \overset{\mathrm{res}}{\longrightarrow} C^{\perp}(N - M)^* \longrightarrow 0. \qquad (4.8)$$

Apply the duality functor $* = \mathrm{Hom}_R(-, R)$ and let $M \subseteq N$ such that

$$|M| = \left\lfloor \frac{d_G - 1}{A} \right\rfloor.$$

Since $C(M)^* = 0$ and $V(M)^* \cong R^n(M)$, the exact sequence in Eq. (4.8) gives the following short exact sequence:

$$0 \longrightarrow C^{\perp}(N - M) \longrightarrow C^{\perp} \longrightarrow V(M) \longrightarrow 0.$$

Since $V(M) \cong R^{|M|}$ is a projective module, the above short exact sequence is split, namely

$$C^{\perp} \cong C^{\perp}(N - M) \oplus R^n(M).$$

Therefore we have

$$\text{f-rank}(C^{\perp}) \geq \text{f-rank}(R^n(M)) = |M| = \left\lfloor \frac{d_G - 1}{A} \right\rfloor.$$

Hence the theorem follows from Lemma 4.1. $\qquad\square$

This leads naturally to the following corollary given the maximum value for each weight.

Corollary 4.1 *Let R be a commutative ring of order 4 and let C be a linear code of length n over R with minimum Hamming weight d_H, minimum Lee weight d_L, minimum Euclidean weight d_E, and minimum Bachoc weight d_B. Then we have*

$$\left\lfloor \frac{d_L - 1}{2} \right\rfloor \leq n - \text{rank}(C), \tag{4.9}$$

$$\left\lfloor \frac{d_E - 1}{4} \right\rfloor \leq n - \text{rank}(C), \tag{4.10}$$

$$\left\lfloor \frac{d_B - 1}{2} \right\rfloor \leq n - \text{rank}(C), \tag{4.11}$$

and

$$d_H - 1 \leq n - \text{rank}(C). \tag{4.12}$$

We generalize the definition to MDS codes to the following. Let R be a commutative ring of order 4 and let C be a linear code over R.

- A code over R meeting bound (4.9) is a Maximum Lee Distance with respect to Rank (MLDR) Code.
- A code over R meeting bound (4.10) is a Maximum Euclidean Distance with respect to Rank (MEDR) Code.

- A code over R meeting bound (4.11) is a Maximum Bachoc Distance with respect to Rank (MBDR) Code.
- A code over R meeting bound (4.12) is a Maximum Hamming Distance with respect to Rank (MHDR) Code.

4.2 Ranks and Kernels of Quaternary Codes

Binary codes, which are the images of linear codes, are not necessarily linear; however, these codes do have some natural structure. In [6, 28], it was shown that a translation invariant propelinear binary code with a commutative group structure must be isomorphic to $\mathbb{Z}_4^\alpha \times \mathbb{Z}_2^\beta$ for some α, β. In fact, it was stated in [6], that the class of additive binary codes considered in that paper coincides with class of additive propelinear codes investigated by Rifà and Pujol in [28]. It is possible that there is also a non-commutative group structure to the translation invariant propelinear code, in which case the structure comes from the quaternion group of order 8, see [28]. The theory for the case when the code is not translation invariant has not yet been developed. However, the following theory applies to arbitrary binary codes so no assumption is made about the structure of the possibly non-linear binary code. To understand this structure, we define the kernel and rank of a binary code. In general, we are concerned with these codes when the binary code is the image of a code under the Gray map, but they apply universally. When the code is the image of a quaternary code we also define quaternary codes associated with these binary codes. We follow the notation given in [7]. We begin with the definition of the kernel of a code which first appears in [3].

Definition 4.3 Let C be a binary code, then $ker(C) = \{\mathbf{v} \in C \mid \mathbf{v} + C = C\}$. If D is a quaternary code, then its kernel is defined to be $\mathcal{K}(D) = \{\mathbf{v} \in D \mid \phi_4(\mathbf{v}) \in ker(\phi_4(D))\}$.

We now examine the code formed by taking the minimal linear code containing the binary code C and the quaternary code which is the preimage of the code.

Definition 4.4 Let C be a binary code. Let $rank(C) = dim(\langle C \rangle)$. Let D be a quaternary code. We let $\mathcal{R}(D) = \{\mathbf{v} \mid \phi_4(\mathbf{v}) \in \langle \phi_4(D) \rangle\}$.

The following results appear first in [25].

Theorem 4.4 *Let C be a binary code containing the all zero vector. Then $ker(C)$ is the intersection of all maximal linear subspaces of C.*

Proof Let C be a binary code containing the all zero vector. If $\mathbf{v} \in ker(C)$, then $\mathbf{v} + \mathbf{0} \in C$ so $ker(C) \subseteq C$. Next, if $\mathbf{v}, \mathbf{w} \in ker(C)$, then $\mathbf{v} + (\mathbf{w} + C) = \mathbf{v} + C = C$ and so $ker(C)$ is a linear code. Therefore we have that $ker(C)$ is a linear code contained in C.

Let D be a maximal linear subspace of C and let $\mathbf{v} \in ker(C)$. Then $\langle D, \mathbf{v} \rangle$ is linear by definition and contained in C since $\mathbf{v} + D \subseteq C$. Since D is maximal we have that $\langle D, \mathbf{v} \rangle = D$ and therefore $\mathbf{v} \in D$. This gives that $ker(C) \subseteq D$ and that $ker(C)$ is contained in the intersection of all maximal linear subspaces of C.

Next, assume $\mathbf{v}$ is in the intersection of all maximal linear subspaces of C and let $\mathbf{w} \in C$. Since $\{\mathbf{0}, \mathbf{w}\}$ is a linear subspace of C it is contained in a maximal linear subspace D. Then $\mathbf{v} \in D$ and so $\mathbf{v} + \mathbf{w} \in D \subseteq C$. This gives that $\mathbf{v} \in ker(C)$ since $\mathbf{v} + \mathbf{w} \in C$ for all $\mathbf{w} \in C$. Therefore, the intersection of all maximal linear subspaces of C is contained in $ker(C)$ and hence the two sets are equal. $\square$

This theorem leads naturally to the following corollary, which was first proved in [3].

Corollary 4.2 *Let C be a binary code containing the all zero vector. Then C is the union of disjoint cosets of the kernel.*

Proof Let C be a binary code. We know that $ker(C)$ is a linear subspace of C. Then noting that if $\mathbf{v} \in ker(C)$ and $\mathbf{w} \in C$, we have that $\mathbf{v} + \mathbf{w} \in C$. We see that $\mathbf{w} + ker(C) \subseteq C$ for all $\mathbf{w} \in C$ which gives the result. $\square$

We note that the binary codes we are most interested in are the image of linear quaternary codes under the Gray map. Therefore, the assumption that the binary code contains the all zero vector is not much of a restriction since the image of a linear quaternary code always contains the all zero vector. We required this restriction to ensure that the kernel be contained in the code.

For vectors $\mathbf{v}, \mathbf{w} \in \mathbb{Z}_4^n$, define $\mathbf{v} * \mathbf{w} = (v_1 w_1, v_2 w_2, \ldots, v_n w_n)$. Namely it is the componentwise product of the two vectors. The following lemma can be proved by simply evaluating the possible cases for the elements of $\mathbb{Z}_4$. It appears first in [20].

Lemma 4.2 *Let $\mathbf{v}, \mathbf{w}$ be vectors in $\mathbb{Z}_4^n$. Then*

$$\phi_4(\mathbf{v} + \mathbf{w}) = \phi_4(\mathbf{v}) + \phi_4(\mathbf{w}) + \phi_4(2\mathbf{v} * \mathbf{w}). \qquad (4.13)$$

The following lemma appears in [17, 18].

Lemma 4.3 *Let C be a linear quaternary code and let $\mathbf{v} \in C$. Then $\mathbf{v} \in \mathcal{K}(C)$ if and only if $2\mathbf{v} * \mathbf{w} \in C$ for all $\mathbf{w} \in C$.*

Proof Assume $\mathbf{v} \in \mathcal{K}(C)$, then $\phi_4(\mathbf{v}) + \phi_4(\mathbf{w}) \in \phi_4(C)$ for all $\mathbf{w} \in C$. Since $\mathbf{v} + \mathbf{w} \in C$ it is linear and therefore $\phi_4(\mathbf{v} + \mathbf{w}) \in \phi(C)$. This gives that $\phi_4(\mathbf{v} + \mathbf{w}) - \phi_4(\mathbf{v}) - \phi_4(\mathbf{w}) = \phi_4(2\mathbf{v} * \mathbf{w}) \in \phi_4(C)$ which gives that $2\mathbf{v} * \mathbf{w} \in C$.

If $\mathbf{v} \in C$ and $2\mathbf{v} * \mathbf{w} \in C$ for all $\mathbf{w} \in C$, then $\phi_4(\mathbf{v} + \mathbf{w}) - \phi_4(2\mathbf{v} * \mathbf{w}) = \phi_4(\mathbf{v}) + \phi_4(\mathbf{w}) \in \phi_4(C)$ and $\mathbf{v} \in \mathcal{K}(C)$. $\square$

The following lemma first appears in [17, 18].

Lemma 4.4 *Let C be a linear quaternary code. Then $\mathcal{K}(C)$ and $\mathcal{R}(C)$ are linear codes.*

Proof Assume $\mathbf{v}, \mathbf{w} \in \mathcal{K}(C)$. This implies that $\phi_4(\mathbf{v}) + \phi_4(\mathbf{w}) \in ker(\phi_4(C))$. By Lemma 4.3, we have $2\mathbf{v} * \mathbf{w} \in C$. This implies that $\phi_4(v + w) = \phi_4(2\mathbf{v} * \mathbf{w}) + \phi_4(v) + \phi_4(\mathbf{w}) \in ker(\phi_4(C))$. Then we have that $\mathbf{v} + \mathbf{w} \in \mathcal{K}(C)$. Therefore, $\mathcal{K}(C)$ is a linear code.

For the second statement, notice that $ker(\langle\phi_4(C)\rangle) = \langle\phi_4(C)\rangle$ since $\langle\phi_4(C)\rangle$ is a linear code. Therefore, $\mathcal{R}(C) = \mathcal{K}(\mathcal{R}(C))$ and hence is a linear code.   □

The previous lemmas give the following.

Theorem 4.5 *Let C be a linear quaternary code. Then $\mathcal{K}(C)$ and $\mathcal{R}(C)$ are linear codes and*

$$\mathcal{K}(C) \subseteq C \subseteq \mathcal{R}(C). \tag{4.14}$$

In [7], it is shown that for cyclic quaternary codes both $\mathcal{K}(C)$ and $\mathcal{R}(C)$ are cyclic codes as well.

The code $\phi_4(C)$ is then a possibly non-linear code which sits between two linear codes. The kernel, in some sense, indicates how non-linear the code is. That is, a very small kernel means that the code is highly non-linear where a large kernel indicates that the code is not that far from linearity.

4.3   X-rings

We have seen that there are 4 commutative Frobenius rings of order 4. As a next step, the Frobenius rings of order 16 were studied. There are twelve local Frobenius non-chain rings of order 16, (see [27] for a description of these rings). These rings all have a maximal ideal that can be written as $\mathfrak{m} = \langle u, v \rangle$ for some pair of elements u, v. This ideal has 8 elements and the minimal ideal $\mathfrak{m}^\perp$ consists of two elements 0 and w for some element w. The remaining ideals are of size 4 and are $\langle u \rangle$, $\langle v \rangle$, and $\langle u + v \rangle$. It follows that the Jacobson radical is $\mathfrak{m}$ and the socle is $\mathfrak{m}^\perp$.

This structure allows for all elements in these local rings to be written as $a + bu + cv + dw$, where $a, b, c, d \in \mathbb{F}_2$. Note that we are not assuming that the ring has characteristic 2. For example, we can have the ring $\mathbb{Z}_4[x]/\langle x^2 \rangle$. Then the maximal ideal is $\mathfrak{m} = \langle 2, x \rangle$, the minimal ideal is $\langle 2x \rangle$, and the characteristic of the ring is 4.

Given the form of these rings, the Gray map, as described for $\mathbb{Z}_4$ and $\mathbb{F}_2 + u\mathbb{F}_2$, can be applied recursively to obtain a Gray map for all of these rings, namely $\phi_{16} : R \rightarrow \mathbb{F}_2^4$ by

$$\phi_{16}(a + bu + cv + dw) = (d, c + d, b + d, a + b + c + d), \tag{4.15}$$

where $a, b, c, d \in \{0, 1\}$.

The form of these rings prompted the following definition first made in [12]. Specifically, we want a family of rings with a common structure and a canonical Gray map which allows us to generalize the results about codes over rings of order 4 and 16.

Definition 4.5 A ring R is an X-ring if it is a finite commutative Frobenius local ring with maximal ideal $\mathfrak{m} = \langle u_1, u_2, \ldots, u_k \rangle$ such that $|R| = 2^{2^k}$ and $|\mathfrak{m}| = \frac{|R|}{2}$, where $u_A = \prod_{i \in A} u_i \neq 0$ for all $A \subseteq \{1, 2, \ldots, k\}$.

Note that in the 12 rings of order 16 it is possible that $uv = 0$. For X-rings we are assuming that this is not possible. Let A be a subset of $\{1, 2, \ldots, k\}$ and let $u_A = \prod_{i \in A} u_i$. It is clear that any element of an X-ring can be written in the form

$$\sum_{A \subseteq \mathcal{P}(\{1,2,\ldots,k\})} \alpha_A u_A,$$

where $\alpha_A \in \mathbb{F}_2$.

Lemma 4.5 *An element $\sum_{A \subseteq \mathcal{P}(\{1,2,\ldots,k\})} \alpha_A u_A$ in an X-ring R is a unit if and only if $\alpha_\emptyset = 1$.*

Proof The result follows from the fact that $\mathfrak{m}$ has 2 cosets in R and that $u_A \in \mathfrak{m}$ whenever A is non-empty. □

Example 4.1 We shall give two examples of X-rings. The first example is the ring $\mathbb{Z}_{2^s}[x]/\langle x^t \rangle$, which has characteristic 2^s, where $s = 2^l$ and $l \geq 0$, $t = 2^{k-l}$. It has maximal ideal $\mathfrak{m} = \langle 2, 4, 16, \ldots, 2^{2^{l-1}}, x, x^2, x^4, \ldots, x^{2^{k-l-1}} \rangle$ and $Soc(\mathbb{Z}_{2^s}[x]/\langle x^t \rangle) = \langle 2^{2^{l-1}} x^{2^{k-l}-1} \rangle$. The second example is the ring $\mathbb{Z}_{2^s}[x]/\langle x^t - 2^r x^m \rangle$ which has characteristic 2^s where $s = 2^l$, $l \geq 0$, $t = 2^{k-l}$, $m < t$, $r \geq 0$. It has maximal ideal $\mathfrak{m} = \langle 2, 4, 16, \ldots, 2^{2^{l-1}}, x, x^2, x^4, \ldots, x^{2^{k-l-1}} \rangle$ and $Soc(\mathbb{Z}_{2^s}/\langle x^t - 2^r x^m \rangle) = \langle 2^{2^{l-1}} x^{2^{k-l}-1} \rangle$.

We define a Gray map for X-rings of order 2^{2^k} recursively. Let ϕ_1 be the map defined on a ring of order 4. That is $\phi_1(a + bu_1) = (b, a + b)$. Take an element c written as $c = c_1 + u_i c_2$, where c_1 and c_2 are elements of the X-ring of order $2^{2^{l-1}}$. Define

$$\phi_i(c) = (\phi_{i-1}(c_2), \phi_{i-1}(c_1) + \phi_{i-1}(c_2)). \tag{4.16}$$

Then extend the map coordinatewise to R^n.

We shall define another Gray map also defined for X-rings. This map will turn out to be conjugate to the previously defined Gray map. We keep two maps because often it is much easier to find results using one of the maps as opposed to the other.

Let R be an X-ring with $|R| = 2^{2^k}$, $\mathfrak{m} = \langle u_1, u_2, \ldots, u_k \rangle$. We define the map $\psi_k : R \to \mathbb{F}_2^{2^k}$. View R as a vector space over $\mathbb{F}_2$ with basis $\{u_A : A \subseteq \{1, 2, \ldots, k\}\}$ where $u_A = \prod_{i \in A} u_i$. Define $\psi_k(u_A) = (c_B)$ where

$$(c_B) = \begin{cases} 1 & B \subseteq A \\ 0 & otherwise. \end{cases}$$

Extend ψ_k linearly to be defined over the entire ring R. Specifically,

$$\psi_k(u_A + u_B) = \psi_k(u_A) + \psi_k(u_B). \qquad (4.17)$$

As an example, if $k = 3$, the subsets are ordered as

$$\emptyset, \{1\}, \{2\}, \{3\}, \{1, 2\}, \{1, 3\}, \{2, 3\}, \{1, 2, 3\}.$$

This gives that $\psi_k(u_1 u_2) = (1\,1\,1\,0\,1\,0\,0\,0)$ and $\psi_k(u_2) = (1\,0\,1\,0\,0\,0\,0\,0)$. Then by linearity, we have $\psi_k(u_2 + u_1 u_2) = (0\,1\,0\,0\,1\,0\,0\,0)$.

It follows from the definition, that the weight of the image an element $c \in R$, $\psi_k(c)$, is odd if and only if c is a unit. Moreover, we have that $\psi_k(u_i)$ has weight 2 for all i, $1 \leq i \leq k$. The following lemma appears in [12].

Lemma 4.6 *The map ψ_k is conjugate to ϕ_k by the permutation*

$$(1, 2^k)(2, 2^k - 1)(3, 2^k - 2)\ldots(2^{k-1}, 2^{k-1} + 1) = \prod_{i=1}^{2^{k-1}} (i, 2^k + 1 - i).$$

Proof For $k = 2$, it is a simple computation to show that they are conjugate via the permutation $(1, 4)(2, 3)$. Then the recursion gives the rest. $\qquad \square$

The following theorem is easily proven from the definition of the maps.

Theorem 4.6 *Let R be an X-ring. The Gray maps ϕ_k and ψ_k are linear if and only if $char(R) = 2$.*

We shall now give three important families of X-rings. These three rings are of particular importance as X-rings since for these rings we have $\phi_k(C^{\perp}) = \phi_k(C)^{\perp}$. We shall prove this result later. This result is not true for all X-rings and we shall give examples of this as well.

- The first family of rings is denoted by R_k and has characteristic 2. These rings have been studied extensively in [14–16, 24]. Define the rings as

$$R_k = \mathbb{F}_2[u_1, u_2, \ldots, u_k]/\langle u_i{}^2, u_i u_j - u_j u_i \rangle. \qquad (4.18)$$

These rings are not chain rings when $k > 1$ and they have characteristic 2. The maximal ideal for the ring R_k is $\mathfrak{m} = \langle u_1, u_2, \ldots, u_k \rangle$. The socle for the ring R_k is $Soc(R_k) = \langle u_1 u_2 \cdots u_k \rangle$. We have that $|R_k| = 2^{2^k}$.

- The second family of rings is denoted by S_k and has characteristic 4. The ring S_1 was first studied in [27]. Define the rings as

$$S_k = \mathbb{Z}_4[u_1, u_2, \ldots, u_k]/\langle u_1^2 - 2u_1, u_2^2 - 2u_2, \ldots, u_k^2 - 2u_k, u_i u_j - u_j u_i \rangle. \qquad (4.19)$$

The maximal ideal for the ring S_k is $\langle 2, u_1, u_2, \ldots, u_k \rangle$. The socle for the ring S_k is $Soc(S_k) = \langle 2u_1 u_2 \cdots u_k \rangle$. We have that $|S_k| = 4^{2^k}$.

- The third family of rings is denoted by T_k and has characteristic 4. The ring T_1 was first studied in [27].

$$T_k = \mathbb{Z}_4[u_1, u_2, \ldots, u_k]/\langle u_1^2 - 2, u_2^2 - 2, \ldots, u_k^2 - 2, u_i u_j - u_j u_i \rangle. \quad (4.20)$$

The maximal ideal for the ring T_k is $\langle 2, u_1, u_2, \ldots, u_k \rangle$. The socle for the ring T_k is $\langle 2u_1 u_2 \cdots u_k \rangle$. We have that $|T_k| = 4^{2^k}$.

The rings R_k, S_k, and T_k are all X-rings. Therefore, they have the already defined Gray map. However, for S_k and T_k there are $k + 1$ generators of the maximal ideal, not k as the notation may seem to indicate. Namely, the generators are $2, u_1, u_2, \ldots, u_k$. Hence we shall use ψ_k for these rings to denote the Gray map defined by recursion on the u_i, which maps to $\mathbb{Z}_4^{2^k}$, and reserve ϕ_{k+1} for the full Gray map to $\mathbb{F}_2^{2^{k+1}}$.

The following results were first shown in [11].

Lemma 4.7 *Let C be a code over R_k, S_k or T_k and let ψ_k be its corresponding Gray map to $\mathbb{Z}_4^{2^k n}$. Then $\psi_k(C^\perp) = \psi_k(C)^\perp$.*

Proof We shall prove the theorem by induction for T_k; the proofs for R_k and S_k are similar. If $k = 1$, we shall show that the Gray images of orthogonal vectors in S_1 are orthogonal in $\mathbb{Z}_4$. Let $\mathbf{v}_1 + u_1 \mathbf{w}_1$ and $\mathbf{v}_2 + u_1 \mathbf{w}_2$ be two orthogonal vectors in S_1, where $\mathbf{v}_i, \mathbf{w}_i$,, are vectors in $\mathbb{Z}_4^n$. This gives

$$[\mathbf{v}_1 + u_1 \mathbf{w}_1, \mathbf{v}_2 + u_1 \mathbf{w}_2] = [\mathbf{v}_1, \mathbf{v}_2] + 2[\mathbf{w}_1, \mathbf{w}_2] + [\mathbf{v}_1, \mathbf{w}_2] + [\mathbf{v}_2, \mathbf{w}_1] = 0.$$

The images of the vectors have the following inner-product:

$$[\psi_1(\mathbf{v}_1 + u_1 \mathbf{w}_1), \psi_1(\mathbf{v}_2 + u_1 \mathbf{w}_2)] = [(\mathbf{w}_1, \mathbf{v}_1 + \mathbf{w}_1), (\mathbf{w}_2, \mathbf{v}_1 + \mathbf{w}_2)]$$
$$= [\mathbf{w}_1, \mathbf{w}_2] + [\mathbf{v}_1, \mathbf{v}_2] + [\mathbf{v}_1, \mathbf{w}_2] + [\mathbf{v}_1, \mathbf{w}_2] + [\mathbf{w}_1, \mathbf{w}_2]$$
$$= 0.$$

This gives that $\phi_1(C^\perp) \subseteq \phi_1(C)^\perp$. The fact that ϕ_1 is a bijection gives the equality of the two sets.

Next, let $\mathbf{v}_1 + u_k \mathbf{w}_1$ and $\mathbf{v}_2 + u_k \mathbf{w}_2$ be two orthogonal vectors in T_{k-1}, where $\mathbf{v}_i, \mathbf{w}_i$ are vectors in T_{k-1}^n. This gives

$$[\mathbf{v}_1 + u_k \mathbf{w}_1, \mathbf{v}_2 + u_k \mathbf{w}_2] = [\mathbf{v}_1, \mathbf{v}_2] + 2[\mathbf{w}_1, \mathbf{w}_2] + [\mathbf{v}_1, \mathbf{w}_2] + [\mathbf{v}_2, \mathbf{w}_1] = 0.$$

The images of the vectors have the following inner-product:

$$[\psi_k(\mathbf{v}_1 + u_k \mathbf{w}_1), \psi_k(\mathbf{v}_2 + u_k \mathbf{w}_2)] = [(\mathbf{w}_1, \mathbf{v}_1 + \mathbf{w}_1), (\mathbf{w}_2, \mathbf{v}_1 + \mathbf{w}_2)$$
$$= [\mathbf{w}_1, \mathbf{w}_2] + [\mathbf{v}_1, \mathbf{v}_2] + [\mathbf{v}_1, \mathbf{w}_2] + [\mathbf{v}_1, \mathbf{w}_2] + [\mathbf{w}_1, \mathbf{w}_2]$$
$$= 0.$$

This gives that $\phi_k(C^\perp) \subseteq \phi_k(C)^\perp$. The fact that ϕ_1 is a bijection gives the equality of the two sets.

Theorem 4.7 *Let C be a self-dual code over S_k or T_k of length n. Then $\psi_k(C)$ is a self-dual code of length $2^k n$ over $\mathbb{Z}_4$.*

Proof Let C be a self-dual code. Then $C = C^\perp$ which gives $\psi_k(C) = \psi_k(C^\perp) = \psi_k(C)^\perp$ by Lemma 4.7.   □

With a similar proof to the proof of Lemma 4.7, we have the following theorem.

Theorem 4.8 *Let C be a code over R_k and let ϕ_k be its corresponding Gray map to $\mathbb{F}_2^{2^k n}$. Then $\psi_k(C^\perp) = \psi_k(C)^\perp$.*

4.4 The Ring $R_{q,\Delta}$

We can generalize the ring R_k into a larger family of rings. These rings have a construction and a Gray map which will make them very useful in the construction of quasicyclic codes. Specifically, cyclic codes over members of this family can be used to give an algebraic understanding of q-ary quasicyclic codes of arbitrary index, see Theorem 6.13 for a complete explanation. This idea was first described in the binary case in [8].

Let $p_1, p_2, \ldots, p_t$ be prime numbers with $t \geq 1$ and $p_i \neq p_j$ if $i \neq j$. Set $\Delta = p_1^{k_1} p_2^{k_2} \cdots p_t^{k_t}$. Let $\{u_{p_i,j}\}_{(1 \leq j \leq k_i)}$ be a set of indeterminates. That is, we write Δ in its unique prime factorization. We construct the following ring:

$$R_{q,\Delta} = \mathbb{F}_q[u_{p_1,1}, \ldots, u_{p_1,k_1}, u_{p_2,1} \ldots, u_{p_2,k_2}, \ldots, u_{p_t,k_t}]/\langle u_{p_i,j}^{p_i} \rangle, \qquad (4.21)$$

where the indeterminates $\{u_{p_i,j}\}_{(1 \leq i \leq t, 1 \leq j \leq k_i)}$ commute. For each $\Delta \in \mathbb{N}$, there is a ring in this family of rings.

An indeterminate $u_{p_i,j}$ can have an exponent in the set

$$J_i = \{0, 1, \ldots, p_i - 1\}.$$

Let $\alpha_i \in J_i^{k_i}$ and denote $u_{p_i,1}^{\alpha_i,1} \cdots u_{p_i,k_i}^{\alpha_i,k_i}$ by $u_i^{\alpha_i}$. For a monomial $u_1^{\alpha_1} \cdots u_t^{\alpha_t}$ in $R_{q,\Delta}$ write u^α, where $\alpha = (\alpha_1, \ldots, \alpha_t) \in J_1^{k_1} \times \cdots \times J_t^{k_t}$. Let

$$J = J_1^{k_1} \times \cdots \times J_t^{k_t}.$$

For an element c in $R_{q,\Delta}$, we can write c as

$$c = \sum_{\alpha \in J} c_\alpha u^\alpha = \sum_{\alpha \in J} c_\alpha u_{p_1,1}^{\alpha_1,1} \cdots u_{p_1,k_1}^{\alpha_1,k_1} \cdots u_{p_t,1}^{\alpha_t,1} \cdots u_{p_t,k_t}^{\alpha_t,k_t}, \qquad (4.22)$$

with $c_\alpha \in \mathbb{F}_q$.

It follows immediately that the ring $R_{q,\Delta}$ is a commutative ring with $|R_{q,\Delta}| = q^{p_1^{k_1} p_2^{k_2} \cdots p_t^{k_t}}$.

Define the ideal $\mathfrak{m} = \langle u_{p_i,j} \rangle_{(1 \leq i \leq t, 1 \leq j \leq k_i)}$. Every element in $R_{q,\Delta}$ can be written as $R_{q,\Delta} = \{a_0 + a_1 m \mid a_0, a_1 \in \mathbb{F}_q, m \in \mathfrak{m}\}$.

It is easy to prove that all units in $R_{q,\Delta}$ are elements of the form $a_0 + a_1 m$, with $m \in \mathfrak{m}$ and $a_0 \neq 0$. First, the following is needed.

The Jacobson radical of $R_{q,\Delta}$ is

$$J(R_{q,\Delta}) = \mathfrak{m} = \langle u_{p_i,j} \rangle_{(1 \leq i \leq t, 1 \leq j \leq k_i)}.$$

For the ring $R_{q,\Delta}$ there is a unique minimal ideal. Therefore, the socle of the ring $R_{q,\Delta}$ is

$$Soc(R_{q,\Delta}) = \{0, u_{p_1,1}^{p_1-1} \cdots u_{p_1,k_1}^{p_1-1} \cdots u_{p_t,1}^{p_t-1} \cdots u_{p_t,k_t}^{p_t-1}\}.$$

The socle of $R_{q,\Delta}$ is the annihilator of $\mathfrak{m}$. We have that $R_{q,\Delta}/J(R_{q,\Delta}) = R_{q,\Delta}/\mathfrak{m} \cong \mathbb{F}_q \cong Soc(\mathfrak{m})$ and therefore $R_{q,\Delta}$ is a Frobenius ring.

We will now describe a Gray map for this ring. Consider the elements in R_Δ as binary vectors of length Δ and call this set A_Δ. Order the elements of A_Δ lexicographically and use this ordering to label the coordinate positions of $\mathbb{F}_q^\Delta$. For $a \in A_\Delta$, define $\Psi : R_{q,\Delta} \to \mathbb{F}_q^\Delta$ as follows. For all $b \in A_\Delta$,

$$\Psi(a)_b = \begin{cases} 1 & \text{if } \widehat{b} \subseteq \{\widehat{a} \cup 1\}, \\ 0 & \text{otherwise}, \end{cases}$$

where $\Psi(a)_b$ indicates the coordinate of $\Psi(a)$ corresponding to the position of the element $b \in A_\Delta$ with the defined ordering.

It follows that $\Psi(a)_b$ is 1 if each indeterminate $u_{p_i,j}$ in the monomial b with non-zero exponent is also in the monomial a with the same exponent; in other words $\bar{b}$ is a subset of $\bar{a}$. Then extend Ψ linearly for all elements of $R_{q,\Delta}$. Note that, generally, orthogonality is not preserved by the map Ψ.

The reason this family of rings is designed in this way is so that the image of cyclic codes over this ring will produce quasicyclic codes, see Sect. 6.4 for a complete description.

4.5 Chain Rings and Principal Ideal Rings

We have generalized the ring $\mathbb{F}_2 + u\mathbb{F}_2$ to the family of rings R_k and to $R_{q,\Delta}$. The field of order 4 generalized to finite fields and codes over these fields have been well studied. The remaining rings of order 4 are $\mathbb{Z}_4$ and $\mathbb{F}_2 + v\mathbb{F}_2$.

The ring $\mathbb{Z}_4$ generalizes to the principal ideal rings $\mathbb{Z}_n$. Using the classical Chinese Remainder Theorem these ring are isomorphic to direct products of chain rings of the form $\mathbb{Z}_{p^e}$ with p a prime.

We generalize the ring $\mathbb{F}_2 + v\mathbb{F}_2$ as follows. For integers $k \geq 1$, define

$$A_k = \mathbb{F}_2[v_1, v_2, \ldots, v_k]/\langle v_i^2 - v_i, v_i v_j - v_j v_i\rangle.$$

This family of rings was first described in [4]. We can define the elements of these rings as follows. Let $B \subseteq \{1, 2, \ldots, k\}$ and then define $v_B = \prod_{i \in B} v_i$. We set $v_\emptyset = 1$. Then each element of A_k can be written in the form $\sum_{B \in \mathcal{P}_k} \alpha_B v_B$ where $\alpha_B \in \mathbb{F}_2$, and $\mathcal{P}_k$ is the power set of the set $\{1, 2, 3, \ldots, k\}$. The ring A_k has characteristic 2 and cardinality 2^{2^k}. For any $A, B \subseteq \{1, 2, \ldots, k\}$ we have that $v_A v_B = v_{A \cup B}$. It follows that

$$\sum_{B \in \mathcal{P}_k} \alpha_B v_B \cdot \sum_{C \in \mathcal{P}_k} \beta_C v_C = \sum_{D \in \mathcal{P}_k} (\sum_{B \cup C = D} \alpha_B \beta_C) v_D.$$

It is easy to see that the only unit in the ring A_k is 1.

Theorem 4.9 *In the ring A_k, the ideal $\langle w_1, w_2, \ldots, w_k\rangle$, where $w_i \in \{v_i, 1 + v_i\}$, is a maximal ideal of A_k with cardinality 2^{2^k-1}.*

Proof The ideal $\langle v_1, v_2, \ldots, v_k\rangle$ consists of all elements of the form $\sum \alpha_B v_B$ with $\alpha_\emptyset = 0$. Therefore the cardinality of the ideal is half the cardinality of the ring A_k. The other maximal ideals described above are isomorphic to this ideal. Specifically, given the maximal ideal $\langle a_1, a_2, \ldots, a_k\rangle$, the isomorphism is formed by mapping a_i to v_i. Therefore these maximal ideals all have the same cardinality.

The ideal is a subgroup of the additive group of the ring. Therefore, the cardinality of an ideal must divide the cardinality of the ring. This ideal has cardinality $\frac{|A_k|}{2}$. Thus it is a maximal ideal.

Denote these maximal ideals in Theorem 4.9 by $\mathfrak{m}_i$. There are 2^k such ideals and $\mathfrak{m}_i^e = \mathfrak{m}_i$ for all i and $e \geq 1$ which gives that its index of stability is 1. The direct sum of any two of these ideals is A_k.

Theorem 4.10 *The ring A_k is isomorphic, via the Chinese Remainder Theorem, to $\mathbb{F}_2^{2^k}$. Therefore the ring A_k is a direct product of finite fields and is a principal ideal ring.*

Proof Using Theorem 4.9, we apply the Chinese Remainder Theorem. This gives that $|A_k/\mathfrak{m}_i| = \frac{|A_k|}{|\mathfrak{m}_i|} = \frac{2^{2^k}}{2^{2^k-1}} = 2$. The ideal $\mathfrak{m}_i$ is a maximal ideal of the ring A_k, which gives that $A_k/\mathfrak{m}_i \cong \mathbb{F}_2$ for all i. $\square$

This theorem gives a natural Gray map, namely the inverse of the Chinese Remainder Theorem. This maps A_k to $\mathbb{F}_2^{2^k}$. For this ring, there is a natural involution defined by $\overline{v_i} = 1 + v_i$, which we use for the Hermitian inner-product. It is this involution that gives the main use of this ring, namely to construct skew-cyclic codes as a generalization of the skew-cyclic codes over $\mathbb{F}_2 + v\mathbb{F}_2$.

Example 4.2 Consider the ideal $\langle v_i \rangle$. The ideal $\langle v_i \rangle$ has elements of the form $\sum_A \alpha_a v_A$ where $v_A = \prod_{j \in A} v_j$ and i must be in A. Hence the cardinality of the ideal is $2^{2^{k-1}}$. Then we have $v_i \overline{v_i} = v_i(1 + v_i) = v_i + v_i = 0$. Hence $C \subset C^H$. Then since $|\langle v_i \rangle| = \sqrt{|A_k|}$, we have that $C = C^H$ and the code is a Hermitian self-dual code of length 1.

For chain rings in general there is an additional generalization of the Lee weight, namely the homogeneous weight. It was introduced in [19]. Let R be a finite chain ring where the maximal ideal is generated by γ. Let $|R| = q^e$ with $|R/\langle \gamma \rangle| = q$. Then the homogeneous weight is defined as:

$$wt_{hom}(x) = \begin{cases} 0 & x = 0 \\ q^{e-1} & x \in \langle \gamma \rangle - \{0\} \\ (q-1)q^{e-2} & x \notin \langle \gamma \rangle. \end{cases}$$

Note that this weight is identical to the Lee weight for the chain ring $\mathbb{Z}_4$. For $e = 1$ it is not exactly the Hamming weight since non-zero elements have weight $\frac{q-1}{q}$ rather than 1. This weight has found numerous applications especially in terms of algebraic geometry codes. See [23, 32] for example.

4.6 Generalized Singleton Bound

In this section, we shall give a generalized version of the classical Singleton bound for linear codes that applies for all finite commutative Frobenius rings. We shall follow the proof as is given by Shiromoto in [31]. Shiromoto proves the results for possibly non-commutative quasi-Frobenius rings. The importance of this bound is that the algebraic structure of a code produces a sharper bound than the strictly combinatorial bound. This will allow us to determine when a linear code can possible be an MDS code in terms of its algebraic structure. We begin with some definitions.

Recall that a monomorphism is an injective homomorphism. An epimorphism is a morphism f that satisfies $g \circ f = h \circ f$ implies $g = h$. For a code C over a ring R, let $f - rank(C)$ denote the free rank of C, that is the multiplicity of free R modules in C as direct summands. Note that the free rank and the rank are only equal if the code itself is free, that is, it is isomorphic to a k fold direct sum of R.

For a linear code C over R define $\mathcal{P}(C)$ as the minimum free R module such that there exists an epimorphism from $\mathcal{P}(C)$ to C and $\mathcal{I}(C)$ as the minimum free R module such that there exists a monomorphism from $\mathcal{P}(C)$ to C. Let C^* denote the

R module $Hom_R(C, R)$ where the action is defined by $r\phi : \mathbf{c} \to r\phi(\mathbf{c})$ for all r in R and all $\phi \in Hom_R(C, R)$. There is an isomorphism Ψ between R^n and $(R^n)^*$ defined by $\Psi_\mathbf{v}(\mathbf{v}') = [\mathbf{v}, \mathbf{v}']$. Then we have that $C^\perp$ is the kernel of $\Psi|_C \to (R^n)^*$.

We now give a lemma which combines Lemmas 2 and 3 in [31] in a different setting. We omit the proof which can be found in [31].

Lemma 4.8 *Let C be a code over a finite commutative Frobenius ring of length n. Then*

$$f - rank(\mathcal{P}(C)) = f - rank(\mathcal{I}(C^*)) \tag{4.23}$$

and

$$f - rank(\mathcal{I}(C)) = f - rank(\mathcal{P}(C^*)). \tag{4.24}$$

Moreover,

$$n = f - rank(C^\perp) + f - rank((\mathcal{P}(C^*))$$
$$= f - rank(\mathcal{I}(C^\perp)) + f - rank(C^*).$$

For a subset M of the coordinates of R^n and a linear code C in R^n define $V(M)$ to be $\{\mathbf{v} \in R^n \mid support(\mathbf{v}) \subset M\}$ and $C(M)$ to be $C \cap V(M)$. This leads to the following theorem which is Proposition 4 in [31].

Theorem 4.11 *Let C be a linear code over a finite commutative Frobenius ring R. Let M be a subset of N which is the coordinates of R^n. Then the sequence*

$$0 \to C^\perp(N - M) \xrightarrow{inc} C^\perp \xrightarrow{cut} V(M) \xrightarrow{\Psi} C(M)^* \to 0 \tag{4.25}$$

of R modules is exact.

Proof Of course the inclusion map is a monomorphism. The fact that Ψ is a surjection follows from the fact that the ring R is self-injective since it is a Frobenius ring. The rest follows from a straightforward computation that $Im(cut)$ is the kernel of Ψ.

Theorem 4.12 *Let C be a linear code of length n over a finite commutative Frobenius ring R. Let $k = \min\{\ell \mid$ there exists a monommorphism from C to R^ℓ as R modules $\}$. Then*

$$d_H(C) \leq n - k + 1. \tag{4.26}$$

Proof Let M be a subset of the coordinates N of R^n such that $|M| = d_H(C) - 1$. It follows that $C(M)^* = \mathbf{0}$ which gives the following short exact sequence:

$$0 \to C^\perp(N - M) \to C^\perp \to V(M) \to \mathbf{0}. \tag{4.27}$$

Since the sequence splits we have

$$C^\perp \cong C^\perp(N - M) \oplus V(M). \tag{4.28}$$

It follows that $f - rank(C^{\perp} \geq |M|)$ which gives that

$$d_H(C) \leq n - f - rank(\mathcal{P}(C^*)) + 1 = n - f - rank(\mathcal{I}(C)) + 1, \qquad (4.29)$$

by the results in Lemma 4.8

If a code meets the bound in Eq. 4.26 then we say that the code is a Maximum Distance with respect to Rank (MDR) code.

Corollary 4.3 *Let C be a linear code over a finite commutative Frobenius principal ideal ring. Then $d_H(C) \leq n - rank(C) + 1$.*

Proof In this case min$\{\ell \mid$ there exists a monommorphism from C to R^ℓ as R modules $\}$ is the rank of the code.

From this corollary, we obtain the classical result for codes over fields which says that $d_H(C) \leq n - \dim(C) + 1$.

We note that an MDR code is not necessarily MDS. For example, consider the code $\{0, 2\}$ of length 1 over $\mathbb{Z}_4$. The minimum distance is 1, and the rank is 1 giving, $1 = 1 - 1 + 1$. However, the code is not MDS since $1 \neq 1 - \frac{1}{2} + 1$.

Corollary 4.4 *Let R be a finite commutative Frobenius ring and let C be a code over R. The code C is an MDS code if and only if C is an MDR code and C is free.*

Proof For a code C to be MDS we have that $d_H(C) = n - log_{|R|}(C) + 1$. Then, if the code is MDR but not free, we have that $|C| < |R|^{n-d_H(C)+1}$. Hence by Eq. 4.26 we have the result.

References

1. Bachoc, C.: Application of coding theory to the construction of modular lattices. J. Combin. Theory Ser. A **78**, 92–119 (1997)
2. Bannai, A., Dougherty, S.T., Harada, M., Oura, M.: Type II codes, even unimodular lattices, and invariant rings. IEEE-IT **45**(4), 1194–1205 (1999)
3. Bauer, H., Ganter, B., Hergert, F.: Algebraic techniques for nonlinear codes. Combinatorica **3**(1), 21–33 (1983)
4. Cengellenmis, Y., Dertli, A., Dougherty, S.T.: Codes over an infinite family of rings with a gray map. Des. Codes Cryptogr. **72**(3), 559–580 (2014)
5. Delsarte, P.: An algebraic approach to the association schemes of coding theory. Philips Res. Rep. Suppl. **10** (1973)
6. Delsarte, P., Levenshtein, V.I.: Association Schemes and Coding Theory, IEEE Trans. Inform. Theory **446** (1998)
7. Dougherty, S.T., Fernández-Córdoba, C.: Kernels and ranks of cyclic and negacyclic quaternary codes. Des. Codes Cryptogr. **81**(2), 347–364 (2016)
8. Dougherty, S.T., Fernández-Córdoba, C., Ten-Valls, R.: Quasi-cyclic codes as cyclic codes over a family of local rings. Finite Fields Appl. **40**, 138–149 (2016)
9. Dougherty, S.T., Gaborit, P., Harada, M., Munemasa, A., Solé, P.: Type IV self-dual codes over rings. IEEE Trans. Inform. Theory **45**(7), 2345–2360 (1999)

10. Dougherty, S.T., Harada, M., Gaborit, P., Solé, P.: Type II codes over $\mathbb{F}_2 + u\mathbb{F}_2$. IEEE Trans. Inform. Theory **45**(1), 32–45 (1999)
11. Dougherty, S.T., Kaya, A., Saltürk, E.: Constructions of self-dual codes and formally self-dual codes over rings. Appl. Algebra Eng. Comm. Comput. **27**(5), 435–449 (2016)
12. Dougherty, S.T., Saltürk, E.: Codes over a family of local frobenius rings, gray maps and self-dual codes. Discrete Appl. Math. **217**, 512–524 (2017)
13. Dougherty, S.T., Shiromoto, K.: Maximum distance codes over rings of order 4. IEEE-IT, **47**1 (2001)
14. Dougherty, S.T., Yildiz, B., Karadeniz, S.: Codes over R_k, gray maps and their binary images. Finite Fields Appl. **17**(3), 205–219 (2011)
15. Dougherty, S.T.: Cyclic codes over R_k. Des. Codes Cryptogr. **63**(1), 113–126 (2012)
16. Dougherty, S.T., Yildiz, B., Karadeniz, S.: Self-dual codes over R_k and binary self-dual codes. Eur. J. Pure Appl. Math. **6**(1), 89–106 (2013)
17. Fernández-Córdoba, C., Pujol, J., Villanueva, M.: On rank and kernel of $\mathbb{Z}_4$-linear codes. Lecture Notes in Computer Science, vol. 5228, pp. 46–55. Springer, Heidelberg (2008)
18. Fernández-Córdoba, C., Pujol, J., Villanueva, M.: $\mathbb{Z}_2\mathbb{Z}_4$-linear codes: rank and kernel. Des. Codes Cryptogr. **56**, 43–59 (2010)
19. Greferath, M., Schmidt, S.E.: Gray isometries for finite chain rings and a nonlinear ternary $(36, 3^{12}, 15)$ code. IEEE Trans. Inform. Theory **45**, 2522–2524 (1999)
20. Hammons, A.R., Kumar, P.V., Calderbank, A.R., Sloane, N.J.A., Solé, P.: The $\mathbb{Z}_4$-linearity of Kerdock, Preparata, Goethals and related codes. IEEE Trans. Inf. Theory **40**, 301–319 (1994)
21. Hammons, A.R., Kumar, P.V., Calderbank, A.R., Sloane, N.J.A., Solé, P.: On the apparent duality of the Kerdock and Preparata codes. In: Applied Algebra, Algebraic Algorithms and Error-Correcting Codes (San Juan, PR, 1993), Lecture Notes in Computer Science, vol. 673, pp. 13–24. Springer, Heidelberg (1993)
22. Honold, T., Landjev, I.: Linear codes over finite chain rings. Electron. J. Comb., **7**(R11) (2000)
23. Honold, T.: Characterization of finite Frobenius rings arch. Math. (Basel) **76**, 406–415 (2001)
24. Karadeniz, S., Dougherty, S.T., Yildiz, B.: Constructing formally self-dual codes over R_k. Discrete Appl. Math. **167**, 188–196 (2014)
25. LeVan, M.: Designs and codes, Ph.D. thesis, Auburn University (1995)
26. McDonald, B. R.: Finite Rings with Identity: Pure and Applied Mathematics, vol. 28, Marcel Dekker, Inc., New York (1974)
27. Martinez-Moro, E., Szabo, S.: On codes over local Frobenius non-chain rings of order 16. Noncommutative Rings Appl. Contemp. Math. **634**, 227–243 (2015)
28. Rifà, J., Pujol, J.: Translation-invariant propelinear codes. IEEE Trans. Inform. Theory l. **43**, 590–598 (1997)
29. Shiromoto, K.: Singleton bounds for codes over finite rings. J. Alg. Combin. **12**(1), 95–99 (2000)
30. Shiromoto, K.: A basic exact sequence for the Lee and Euclidean weights of linear codes over $\mathbb{Z}_\ell$. Lin. Alg. Appl. **295**, 191–200 (1999)
31. Shiromoto, K.: A Singleton bound for linear codes over quasi-Frobenius rings. In: AAECC-13, Honolulu (1999)
32. Voloch, J.F., Walker, J.: Homogeneous weights and exponential sums. Finite Fields Appl. **9**(3), 310–321 (2003)

Chapter 5
Self-dual Codes

Self-dual codes are one of the most important classes of codes. They have been widely studied for both codes over fields and codes over rings. There are numerous connections between self-dual codes over rings and fields and combinatorics, design theory, and number theory. For example, one of the most successful techniques for producing interesting designs uses self-dual codes over fields [2] and one of the most powerful techniques for producing optimal unimodular lattices uses self-dual codes over rings [3]. Moreover, the well known proof of the non-existence of the projective plane of order 10 used the theory of binary self-dual codes, see [26] for a complete explanation of this proof. In addition to their numerous applications in mathematics, self-dual codes are interesting in their own rite. In this chapter, we shall show when self-dual codes exist in general for codes over Frobenius rings and then look at various connections to other mathematical objects.

5.1 Self-dual Codes Over Frobenius Rings

Self-dual codes over rings have been a widely studied object; see [31] for a detailed description of the existing literature on self-dual codes. Part of this has been fueled by the relationship between self-dual codes and unimodular lattices; see [8] for a detailed description of this. In this chapter, we shall take a more general approach to self-dual codes over rings rather than handling various rings separately. That is, we shall establish existence of self-dual codes in a very broad sense and then look to particular rings for various applications.

Recall that a code is said to be self-dual if $C = C^\perp$. Under this definition, it is immediate that a self-dual code must be linear since the code $C^\perp$ is always linear. There are other notions of self-duality which we shall address at the end of the chapter.

© The Author(s) 2017
S.T. Dougherty, *Algebraic Coding Theory Over Finite Commutative Rings*,
SpringerBriefs in Mathematics, DOI 10.1007/978-3-319-59806-2_5

The following lemmas are standard tools in determining when self-dual codes exist.

Lemma 5.1 *Let R be a finite commutative Frobenius ring. If $|R|$ is not a square and there exists a self-dual code C of length n, then n must be even.*

Proof We know from Corollary 3.2 that $|C||C^\perp| = |R|$. This gives that $|C| = |R|^{\frac{n}{2}}$. If $|R|$ is not a square, then $|C|$ is not an integer, which is a contradiction. Hence n must be even. $\square$

We can say more if the alphabet is a finite field.

Lemma 5.2 *Let F be a finite field. If C is a self-dual code over F of length n, then n must be even.*

Proof If C is a a self-dual code, then $dim(C) + dim(C^\perp) = 2dim(C) = n$ and therefore n is even. $\square$

Of course, this is not true when the underlying alphabet is not a field. For example, the code $\{0, 2\}$ is a self-dual code of length 1 over $\mathbb{Z}_4$.

We continue with an application of the Chinese Remainder Theorem. This theorem allows us to focus on local rings in order to determine when self-dual codes exist.

The following result was first proven for codes over $\mathbb{Z}_k$ in [15]. It was proven for codes over Frobenius rings in [16].

Theorem 5.1 *Let R be a finite commutative Frobenius ring that is isomorphic via the Chinese Remainder Theorem to $R_1 \times R_2 \times \cdots \times R_s$. Let C_i be a code over R_i and let $C = CRT(C_1, C_2, \ldots, C_s)$. Then C is a self-dual code over R if and only if C_i is a self-dual code over R_i for all i.*

Proof By Theorem 2.7 and Corollary 2.1 we have that $|C| = \prod |C_i|$ and $C^\perp = CRT(C_1^\perp, C_2^\perp, \ldots, C_s^\perp)$.

If $C = C^\perp$ then

$$C = CRT(C_1^\perp, C_2^\perp, \ldots, C_s^\perp)$$

and $C_i = C_i^\perp$. If $C_i = C_i^\perp$ for all i, then

$$C = CRT(C_1, C_2, \ldots, C_s) = CRT(C_1^\perp, C_2^\perp, \ldots, C_s^\perp) = C^\perp.$$

This gives the result. $\square$

Example 5.1 By Theorem 4.10, we have that $A_k = \mathbb{F}_2[v_1, v_2, \ldots, v_k]/\langle v_i^2 = v_i, v_i v_j = v_j v_i \rangle$ is isomorphic via the Chinese Remainder Theorem to $\mathbb{F}_2^{2^k}$. Therefore by Theorem 5.1 we have that there exists a self-dual code over A_k if and only if there exists a binary self-dual code. Specifically, a self-dual code over A_k exists if and only if the length is even.

Lemma 5.3 *Let R be a finite commutative Frobenius ring and let C and D be a self-dual codes of length n and m respectively. Then the direct product $C \times D$ is a self-dual code of length $n + m$ over R.*

Proof Let $(\mathbf{v}, \mathbf{w}), (\mathbf{v}', \mathbf{w}') \in C \times D$. Then

$$[(\mathbf{v}, \mathbf{w}), (\mathbf{v}', \mathbf{w}')] = [\mathbf{v}, \mathbf{v}'] + [\mathbf{w}, \mathbf{w}'] = 0 + 0 = 0.$$

This gives that $C \times D$ is a self-orthogonal code. Then we have that $|C \times D| = |C| \cdot |D| = |R|^{\frac{n}{2}} |R|^{\frac{m}{2}} = |R|^{\frac{n+m}{2}}$. Therefore $C \times D$ is a self-dual code of length $n + m$. $\square$

Moreover, it is immediate that if C and D are free codes, then $C \times D$ is a free code.

Example 5.2 Consider the ring $R_k = \mathbb{F}_2[u_1, u_2, \ldots, u_k]/\langle u_i^2, u_i u_j - u_j u_i \rangle$. Recall that $|R_k| = 2^{2^k}$. The ideal $\langle u_i \rangle$ has elements of the form $\sum_A \alpha_a u_A$ where $u_A = \prod_{j \in A} u_j$ and i must be in A. Hence the cardinality of the ideal is $2^{2^{k-1}}$. We also have that $u_i^2 = 0$, therefore the ideal is self-orthogonal. Then since the cardinality of the ideal is $\sqrt{|R_k|}$ and the ideal is self-orthogonal we have that the ideal is a self-dual code of length 1. Then we can use Lemma 5.3 to obtain that there are self-dual codes of all lengths over R_k.

In general, the key to finding free self-dual codes of even length is to find an element α in the ring whose square is -1. It is well known that a finite field has a square root of -1 if the characteristic of the field is 1 (mod 4). We shall exploit this fact to find such an element in local rings with the property that the ring modded out by its maximal ideal is a field of characteristic 1 (mod 4). We begin with a lifting lemma.

Lemma 5.4 *Let R be a finite local commutative Frobenius ring with maximal ideal $\mathfrak{m}$ such that $R/\mathfrak{m}$ is a field of characteristic p, where p is an odd prime. Let $S_i = R/\mathfrak{m}^i$. If there exists $\alpha \in S_i$ with $\alpha^2 = -1$, then there exists $\beta \in S_{i+1}$ with $\beta^2 = -1$.*

Proof Let $\alpha \in S_i$ with $\alpha^2 = -1$. Let $\beta = \alpha + \gamma_i$ be an element in S_{i+1}, where $\gamma_i + \mathfrak{m}^{i+1} \in \mathfrak{m}^i/\mathfrak{m}^{i+1}$. Then we have that

$$\begin{aligned}
(\alpha + \gamma_i)^2 &\equiv \alpha^2 + 2\alpha\gamma_i + \gamma_i^2 \pmod{\mathfrak{m}^{i+1}} \\
&\equiv \alpha^2 + 2\alpha\gamma_i \pmod{\mathfrak{m}^{i+1}} \\
&\equiv \delta - 1 + 2\alpha\gamma_i \pmod{\mathfrak{m}^{i+1}},
\end{aligned}$$

for some $\delta \in \mathfrak{m}^i$ since $\alpha^2 = -1 \in S_i$.

Next we show that there exists an element γ_i such that $\delta - 1 + 2\alpha\gamma_i = -1 \in S_{i+1}$. We have that

$$\delta - 1 + 2\alpha\gamma_i = -1 \pmod{\mathfrak{m}^{i+1}} \iff \delta = -2\alpha\gamma_i \pmod{\mathfrak{m}^{i+1}}.$$

Since p is odd, 2 is relatively prime to p. Hence the element 2 is a unit. Since $\alpha^2 \equiv -1$ (mod m), this implies that α is a unit in $R/\mathfrak{m}$. Let $\gamma_i = -\delta(2\alpha)^{-1}$. Then $\alpha_i + \mathfrak{m}^{i+1} \in \mathfrak{m}^i/\mathfrak{m}^{i+1}$ and $(\beta)^2 = (\alpha + \gamma_i)^2 \equiv \delta - 1 + 2\alpha\alpha_i \equiv -1$ in S_{i+1} since elements in $\mathfrak{m}^{i+1}$ are 0 in S_{i+1}. $\qquad\square$

Corollary 5.1 *Let R be a finite local commutative Frobenius ring with characteristic congruent to 1 (mod 4). Then there exists an $\alpha \in R$ with $\alpha^2 = -1$.*

Proof The field $R/\mathfrak{m}$ has characteristic 1 (mod 4) and hence has a square root of -1. Then, by induction using Lemma 5.4, we have the result. $\qquad\square$

Notice that this result does not necessarily hold when $R/\mathfrak{m}$ has characteristic 2. For example, $\mathbb{Z}_4$ is a local ring and $\mathbb{Z}_4/\langle 2 \rangle \cong \mathbb{F}_2$, which has a square root of -1, but the ring $\mathbb{Z}_4$ does not.

We can use this result to get the following theorem.

Theorem 5.2 *Let R be a finite local commutative Frobenius ring with characteristic congruent to 1 (mod 4). Then there exists self-dual codes for all even lengths over R.*

Proof By Corollary 5.1, the ring R has an element α with $\alpha^2 = -1$. Then the code generated by $(1, \alpha)$ is a self-dual code of length 2. Then, by applying Lemma 5.3, inductively, we have the result. $\qquad\square$

Lemma 5.5 *Let R be a finite local commutative Frobenius ring with maximal ideal $\mathfrak{m}$ such that $R/\mathfrak{m}$ is a field of characteristic p, where p is an odd prime. Let $S_i = R/\mathfrak{m}^i$. If there exists $\alpha, \beta \in S_i$ with $\alpha^2 + \beta^2 = -1$, then there exists $\gamma, \delta \in S_{i+1}$ with $\gamma^2 + \delta^2 = -1$.*

Proof Let $\alpha, \beta \in S_i$ with $\alpha^2 + \beta^2 = -1$. Let $\gamma = \alpha + \epsilon_i$, $\delta = \beta + \zeta_i$ be elements in S_{i+1}, where $\epsilon_i + \mathfrak{m}^{i+1} \in \mathfrak{m}^i/\mathfrak{m}^{i+1}$ and $\zeta_i + \mathfrak{m}^{i+1} \in \mathfrak{m}^i/\mathfrak{m}^{i+1}$. Then we have that

$$(\alpha + \epsilon_i)^2 + (\beta + \zeta_i)^2 \equiv \alpha^2 + \beta^2 + 2\alpha\epsilon_i + 2\beta\zeta_i + \epsilon_i^2 + \zeta_i^2 \pmod{\mathfrak{m}^{i+1}}$$
$$\equiv \alpha^2 + \beta^2 + 2\alpha\epsilon_i + 2\beta\zeta_i \pmod{\mathfrak{m}^{i+1}}$$
$$\equiv \theta - 1 + 2\alpha\epsilon_i + 2\beta\zeta_i \pmod{\mathfrak{m}^{i+1}}.$$

Next we show that there exist elements ϵ_i, ζ_i such that $\theta - 1 + 2\alpha\epsilon_i + 2\beta\zeta_i = -1 \in S_{i+1}$. We have that

$$-1 = \theta - 1 + 2\alpha\epsilon_i + 2\beta\zeta_i \iff \theta = -2\alpha\epsilon_i - 2\beta\zeta_i \pmod{\mathfrak{m}^{i+1}}.$$

We know at least one of α and β is not 0. Without loss of generality, assume α is not 0. Then take $\zeta_i = 0$ and notice that 2 and α are units we take $\epsilon_i = -\frac{\theta}{2\alpha}$. This gives our desired result. $\qquad\square$

Corollary 5.2 *Let R be a finite local commutative Frobenius ring with characteristic congruent to 1 (mod 4). Then there exists $\alpha, \beta \in R$ with $\alpha^2 + \beta^2 = -1$.*

Proof The field $R/\mathfrak{m}$ has characteristic 3 (mod 4) and hence γ, δ with $\gamma^2 + \delta^2 = -1$. Then by induction using Lemma 5.5 we have the result. $\qquad\square$

This result leads naturally to the following theorem.

Theorem 5.3 *Let R be a finite local commutative Frobenius ring with characteristic congruent to 3 (mod 4). Then there exists self-dual codes for all lengths congruent to 0 (mod 4) over R.*

Proof By Corollary 5.2, the ring R has elements α, β with $\alpha^2 + \beta^2 = -1$. Then the code $C = \langle (1, 0, \alpha, \beta), (0, 1, -\beta, -\alpha) \rangle$ is a self-dual code of length 4. Then, by applying Lemma 5.3 inductively, we have the result. $\qquad\square$

Finally, we can obtain our main result of this section.

Theorem 5.4 *Let R be a finite commutative Frobenius ring that is isomorphic via the Chinese Remainder Theorem to $R_1 \times R_2 \times \cdots \times R_s$, where R_i is a local ring. Then if $\mathfrak{m}_i$ is the maximal ideal of R_i and $R_i/\mathfrak{m}_i$ has characteristic 1 (mod 4), then there exists self-dual codes over R for all even lengths. If there exists i where $R_i/\mathfrak{m}_i$ has characteristic 3 (mod 4), then there exists self-dual codes over R for all lengths congruent to 0 (mod 4).*

Proof The result follows immediately by applying Theorem 5.1 to Theorems 5.2 and 5.3. $\qquad\square$

For chain rings, the situation is easier to handle. In a chain ring, the maximal ideal is $\mathfrak{m} = \langle \gamma \rangle$. Then there exists a minimal positive integer e with $\gamma^e = 0$, which is called the index of niloptency. Every ideal in a chain ring is of the form $\langle \gamma^i \rangle$ for some i. This leads to the following.

Theorem 5.5 *Let R be a finite commutative chain ring with index of nilpotency e and maximal ideal $\mathfrak{m} = \langle \gamma \rangle$. If e is even, then $\mathfrak{a} = \langle \gamma^{\frac{e}{2}} \rangle$ is a self-dual code of length 1.*

Proof We have that $\gamma^{\frac{e}{2}} \gamma^{\frac{e}{2}} = 0$ and so $\mathfrak{a} \subseteq \mathfrak{a}^\perp$. Assume that $\mathfrak{a} \neq \mathfrak{a}^\perp$. Then $\mathfrak{a}^\perp = \langle \gamma^j \rangle$ with $j < \frac{e}{2}$. Then $\gamma^{\frac{e}{2}} \gamma^j = 0$ contradicting that e is minimal. Therefore $\mathfrak{a} = \mathfrak{a}^\perp$ and is a self-dual code of length 1. $\qquad\square$

This leads immediately to the following corollary.

Corollary 5.3 *Let R be a finite commutative chain ring with index of nilpotency e. If e is even, then there exists self-dual codes of length n for all n.*

Proof The result follows immediately by applying Lemma 5.3 to the self-dual code of length 1 in Theorem 5.5. $\qquad\square$

Example 5.3 If k is a positive integer greater than 1, then $\langle k \rangle$ is a self-dual code in $\mathbb{Z}_{k^2}$. Direct products of this code give self-dual codes of all lengths over $\mathbb{Z}_{k^2}$.

If $\mathfrak{a}$ is a self-dual code of length 1 it must satisfy $|\mathfrak{a}|^2 = |R|$. Therefore, it is necessary for $|R|$ to be a square for a self-dual code of length 1 to exist. However, it is not necessary for the ring to be a chain ring. For example, consider the ring of order 16, $\mathbb{Z}_4[x]/\langle x^2 \rangle$. Here $\langle 2 \rangle$, $\langle x \rangle$ and $\langle 2 + x \rangle$ are all self-dual codes of length 1, but the ring is not a chain ring.

Theorem 5.6 *Let R be a finite commutative Frobenius ring of order k^2. If there exists an odd number of ideals of order k, then there exists a self-dual code of length 1.*

Proof Since any ideal in R satisfies $(\mathfrak{a}^\perp)^\perp = \mathfrak{a}$ and $|\mathfrak{a}||\mathfrak{a}^\perp| = |R|$, each ideal of order k is matched with a unique ideal of order k as its orthogonal. Since the number of such ideals is odd, at least one ideal must be its own orthogonal and therefore is a self-dual code of length 1. □

Example 5.4 In the ring of order 16, $\mathbb{Z}_4[x]/\langle x^2 + 2x \rangle$, we have three ideals of order 4. Here $\langle x \rangle^\perp = \langle 2 + x \rangle$. Then the remaining ideal of order 4, namely $\langle 2 \rangle$, is a self-dual code of length 1.

5.2   Connections to Lattices

We shall describe a connection between self-dual codes over rings and unimodular lattices. This has been one of the most productive areas of research for codes over rings since the results obtained by studying codes over rings have, in general, been much stronger than the results obtained by studying codes over finite fields. For example, while the extremal lattice in 24 dimensions can be obtained from a binary code, namely the extended Golay code, the connection will not produce an extremal lattice in 72 dimensions. This is because the maximal minimum norm obtainable from a binary code is 2. However, such an extremal lattice can be obtained from a code over the ring $\mathbb{Z}_4$, see [20] for a description.

We shall give the usual definitions of lattices. For a complete description of lattices, especially in relation to codes, see [8].

Let $\mathbb{F}^n$ be an n-dimensional space where $\mathbb{F}$ is an infinite division ring. We attach the standard Hermitian inner-product, namely $\mathbf{v} \cdot \mathbf{w} = \sum v_i \overline{w_i}$. Notice that we reserve the notation $[\mathbf{v}, \mathbf{w}]$ for codes and use the alternate notation when dealing with lattices. Let $\mathcal{O}$ denote a ring of integers in $\mathbb{F}$. An n-dimensional lattice L in $\mathbb{F}^n$ is a free $\mathbb{Z}$-module spanned by n linearly independent vectors.

The fundamental volume $V(L)$ of L is $|\det G|$, where G is a generator matrix formed by the n linearly independent vectors which generate L.

The dual lattice L^* is given by $L^* = \{\mathbf{v} \in \mathbb{F}^n \mid \mathbf{v} \cdot \mathbf{w} \in \mathbb{Z} \text{ for all } \mathbf{w} \in L\}$. A lattice L is integral if $L \subseteq L^*$ and is unimodular if $L = L^*$. If the norm $\mathbf{v} \cdot \mathbf{v}$ is in $2\mathcal{O}$ for all $\mathbf{v} \in L$, then the lattice is said to be even and it is odd otherwise.

When $\mathbb{F} = \mathbb{R}$, we have that $\mathcal{O} = \mathbb{Z}$ and $\overline{a} = a$. When $\mathbb{F} = \mathbb{C}$, we have that $\mathcal{O} = \mathbb{Z}[i]$ and $\overline{a + bi} = a - bi$. When $\mathbb{F} = \mathbb{H}$, we have that $\mathcal{O} = \mathbb{Z}[i, j, k]$ and $\overline{a + bi + cj + dk} = a - bi - cj - dk$.

The minimum norm of a lattice L is the smallest norm among all nonzero vectors of Λ.

We shall describe three families of rings that will be useful in constructing unimodular lattices. The first is the well known family $\mathbb{Z}_{2k}$. The second is the family

$$\Theta_{2k} = \mathbb{Z}_{2k}[i]/\langle i^2 + 1 \rangle. \tag{5.1}$$

The first ring in this family, Θ_1, is actually isomorphic to $R_1 = \mathbb{F}_2 + u\mathbb{F}_2$ where i corresponds to $1 + u$. This family of rings was first studied in [10]. We attach the standard involution to this family where $\overline{a + bi} = a - bi$. The associated Hermitian inner-product is defined with respect to this involution. That is $[\mathbf{v}, \mathbf{w}]_H = \sum v_i \overline{w_i}$. For this ring, we are not concerned with the Euclidean dual.

The third family is

$$\Sigma_{2k} = \mathbb{Z}_{2k}[i, j, k]/\langle i^2 + 1, j^2 + 1, k^2 + 1, ijk + 1 \rangle. \tag{5.2}$$

This family of rings was first studied in [9]. Note that unlike most of the rings in this text, the ring is not commutative except when $k = 1$. However, we shall describe an inner-product on the ring that makes it act in a way that is very similar to commutative rings.

For Σ_k^n we attach the inner-product $[\mathbf{v}, \mathbf{w}]_H = \sum v_i \overline{w_i}$, where

$$\overline{a + bi + cj + dk} = a - bi - cj - dk.$$

Notice that the usual computation for the quaternions gives that $\overline{\alpha\beta} = \overline{\alpha}\,\overline{\beta}$ and $\overline{\overline{\alpha}} = \alpha$ for all $\alpha, \beta \in \Sigma_k$.

In general, for non-commutative rings, we need both a left and a right orthogonal, however, the following lemma eliminates the need for this.

Lemma 5.6 Let $\mathbf{v}, \mathbf{w} \in \Sigma_k^n$. Then we have that $[\mathbf{v}, \mathbf{w}]_H = 0$ if and only if $[\mathbf{w}, \mathbf{v}]_H = 0$.

Proof If $[\mathbf{v}, \mathbf{w}]_H = 0$, then $\sum v_i \overline{w_i} = 0$. Then $\overline{\sum v_i \overline{w_i}} = \overline{0} = 0$, which gives that $\sum w_i \overline{v_i} = 0$. The other direction is the same computation where the roles of $\mathbf{v}$ and $\mathbf{w}$ are reversed.

It follows from the lemma that we can make a single definition of the orthogonal rather than a left and right diagonal. Namely, we let $C^H = \{\mathbf{v} \in \Sigma_k^n \mid [\mathbf{v}, \mathbf{w}]_H = 0$ for all $\mathbf{w} \in C\}$. For this ring, we are not concerned with the Euclidean dual.

For both Θ_k and Σ_k we define the norm of an element α to be $N(\alpha) = \alpha\overline{\alpha}$. Then $N(\mathbf{v}) = \sum N(v_i)$. For an element $a \in \mathbb{Z}_{2k}$ we define its Euclidean norm to be $\min\{a^2, (2k - a)^2\}$ where the squares are read modulo $4k$.

Originally, Type I and Type II codes were defined for binary self-dual codes. Namely a Type II code was a binary self-dual code where all of the weights were doubly-even and a Type I code was a self-dual code that was not Type II. We shall now generalize this definition to the rings in question.

Definition 5.1 A code over $\mathbb{Z}_{2k}$ is Type II if $N(\mathbf{v}) \equiv 0 \pmod 4$ for all $\mathbf{v} \in C$ and is Type I otherwise. A code over Θ_k is Type II if $N(\mathbf{v}) = 0$ in Θ_{4k} for all $\mathbf{v} \in C$ and is Type I otherwise. A code over Σ_k is Type II if $N(\mathbf{v}) = 0$ in Σ_{4k} for all $\mathbf{v} \in C$ and is Type I otherwise.

Construction A was first described for binary codes; see [8] for a description. It was extended to codes over $\mathbb{Z}_{2K}$ in [3], to codes over Θ_{2k} in [10], and to codes over Σ_{2k} in [9].

Let C be a code over $\mathbb{Z}_{2k}$, Θ_{2k}, or Σ_{2k}. Let $\mathbb{F}$ be one of $\mathbb{R}$, $\mathbb{C}$, or $\mathbb{H}$ and let $\mathcal{O}$ be the corresponding ring of integers. Let ρ_R be a map from R, where R is one of $\mathbb{Z}_{2k}$, Θ_{2k}, or Σ_{2k}, to the corresponding ring of integers sending $0, 1, \ldots, k$ to $0, 1, \ldots, k$ and $k+1, \ldots, 2k-1$ to $1-k, \ldots, -1$, respectively and sending i, j, k to the corresponding elements of the same name in $\mathbb{C}$ and $\mathbb{H}$. Then extend ρ coordinatewise to R^n. Note here that we have carefully chosen this description of these rings to facilitate construction A.

Define the following lattice from a code C:

$$\Lambda_{\mathbb{F}}(C) = \frac{1}{\sqrt{2k}}\{\rho(C) + 2k\mathcal{O}^n\}. \tag{5.3}$$

The following appear in [3, 9, 10]. We denote the minimum Euclidean weight of a code C by $d_E(C)$.

Theorem 5.7 *If C is a self-dual code over $\mathbb{Z}_{2k}$, then $\Lambda_{\mathbb{R}}(C)$ is a real unimodular lattice and $\Lambda_{\mathbb{R}}(C)$ is even if C is Type II. The minimum norm of the lattice is $\min\{\frac{d_E(C)}{2k}, 2k\}$. If C is a self-dual code over Θ_{2m}, then $\Lambda_{\mathbb{C}}(C)$ is a complex unimodular lattice and $\Lambda_{\mathbb{C}}(C)$ is even if C is Type II. The minimum norm of the lattice is $\min\{\frac{d_E(C)}{2k}, 2k\}$. If C is a self-dual code over Σ_{2k}, then $\Lambda_{\mathbb{H}}(C)$ is a Quaternionic unimodular lattice, and $\Lambda_{\mathbb{H}}(C)$ is even if C is Type II. The minimum norm of the lattice is $\min\{\frac{d_E}{2k}, 2k\}$.*

Proof Let n denote the length of the code. We shall prove all three cases at once. Let $\mathbf{v}, \mathbf{w} \in \Lambda_{\mathbb{F}}(C)$. We have that $\mathbf{v} = \frac{1}{\sqrt{2k}}(\mathbf{v}_0 + 2k\mathbf{v}_1)$ and $\mathbf{w} = \frac{1}{\sqrt{2k}}(\mathbf{w}_0 + 2k\mathbf{w}_1)$, where $\mathbf{v}_0, \mathbf{w}_0 \in C$ and $\mathbf{w}_0, \mathbf{w}_1 \in 2k\mathcal{O}$. It follows that

$$\mathbf{v} \cdot \mathbf{w} = \frac{1}{2k}(\mathbf{v}_0 \cdot \mathbf{w}_0 + 2k\mathbf{v}_0 \cdot \mathbf{w}_1 + 2k\mathbf{v}_1 \cdot \mathbf{w}_0 + 4k^2\mathbf{v}_1 \cdot \mathbf{w}_1. \tag{5.4}$$

Then since $[\mathbf{v}_0, \mathbf{w}_0] = 0$ in the rings, this implies that $\mathbf{v}_0 \cdot \mathbf{w}_0 \in 2k\mathcal{O}$ and we have that $\mathbf{v} \cdot \mathbf{w} \in \mathcal{O}$. Hence the lattice is integral.

Then we have that $V(2k\mathcal{O}^n) = (2k)^n$ and $|\sqrt{2k}\Lambda(C)/2k\mathcal{O}^n| = (2k)^{\frac{n}{2}}$. This gives that $V(\sqrt{2k}\Lambda(C)) = (2k)^{\frac{n}{2}}$, which gives that $V(\Lambda(C)) = 1$. Therefore, the lattice is integral with volume 1 and hence is unimodular.

If C is Type II, then we have that

$$
\begin{aligned}
\mathbf{v} \cdot \mathbf{v} &= \frac{1}{2k}(\mathbf{v}_0 \cdot \mathbf{v}_0 + 2k\mathbf{v}_0 \cdot \mathbf{v}_1 + 2k\mathbf{v}_1 \cdot \mathbf{v}_0 + 4k^2\mathbf{v}_1 \cdot \mathbf{v}_1) \\
&= \frac{1}{2k}(\mathbf{v}_0 \cdot \mathbf{v}_0 + 2k(\mathbf{v}_0 \cdot \mathbf{v}_1 + \overline{\mathbf{v}_0 \cdot \mathbf{v}_1} + 4k^2\mathbf{v}_1 \cdot \mathbf{v}_1) \\
&= \frac{1}{2k}(\mathbf{v}_0 \cdot \mathbf{v}_0 + 2k(2r) + 4k^2\mathbf{v}_1 \cdot \mathbf{v}_1).
\end{aligned}
$$

Then since $N(\mathbf{v}_0)$ is 0 modulo $4k$, we have that this norm is in $2\mathcal{O}$. Therefore, if the code is Type II, then the lattice is Type II as well.

Finally, if $\mathbf{v} = \mathbf{v}_0 + 2k\mathbf{v}_1$ as before, it is immediate that $\mathbf{v} \cdot \mathbf{v} \geq \frac{1}{\sqrt{2k}}\mathbf{v}_0 \cdot \frac{1}{\sqrt{2k}}\mathbf{v}_0$ which gives that minimum norm is $\min\{\frac{d_E}{2k}, 2k\}$. $\square$

We have seen that a quaternary code can be used to construct a real extremal lattice in $\mathbb{R}^{72}$ and a binary code can be used to construct a real extremal lattice in $\mathbb{R}^{24}$. In general, we would like to know exactly when we are able to do this.

Question 5.1 This question has three parts.

- Determine for which k and n can a real extremal lattice in $\mathbb{R}^n$ be constructed from a code over $\mathbb{Z}_{2k}$ using the map $\Lambda_{\mathbb{R}}$.
- Determine for which k and n can a complex extremal lattice in $\mathbb{C}^n$ be constructed from a code over Θ_{2k} using the map $\Lambda_{\mathbb{C}}$.
- Determine for which k and n can a quaternionic extremal lattice in $\mathbb{H}^n$ be constructed from a code over Σ_{2k} using the map $\Lambda_{\mathbb{H}}$.

5.3 Connections to Binary Self-dual Codes

In this section, we begin by giving a very brief explanation of the relation between invariant theory and self-dual codes. The classical relationship has been well covered in [27] and much expanded connection has been explained in [30]. We only include here what we need in order to describe some open problems.

If C is a self-dual code over a ring R with $|R| = s$, then the weight enumerator is held invariant by the MacWilliams relations and hence by the following matrix:

$$
M = \frac{1}{\sqrt{s}}\begin{pmatrix} 1 & s-1 \\ 1 & -1 \end{pmatrix}.
$$

This is because the weight enumerator of the code is the same as the weight enumerator of the orthogonal code. In other words, M sends (x, y) to $\frac{1}{\sqrt{s}}(x + (s-1)y, x - y)$ which is precisely the action of the MacWilliams relations. Hence, if $W_C(x, y)$ is the weight enumerator of a self-dual code, then acting with this matrix will give $W_C(x, y)$.

We can then apply the following famous theorem of Molien, see [27].

Theorem 5.8 (Molien) *Define the series $\Phi(\lambda) = \sum a_i \lambda^i$ for a group G, where there are a_i independent polynomials held invariant by the group. For any finite group G of complex m by m matrices, $\Phi(\lambda)$ is given by*

$$\Phi(\lambda) = \frac{1}{|G|} \sum_{A \in G} \frac{1}{det(I - \lambda A)}, \tag{5.5}$$

where I is the identity matrix.

The invariants for this group were found in Chap. 19 of [27] long before anyone was interested in codes over rings. They are the following.

If C is a self-dual code over a ring of size s, then the weight enumerator is held invariant by the matrix corresponding to the MacWilliams relations

$$\begin{pmatrix} 1 & s-1 \\ 1 & -1 \end{pmatrix}.$$

If s is not a square, then it is also held invariant by

$$\begin{pmatrix} -1 & 0 \\ 0 & -1 \end{pmatrix}.$$

Then it follows that the weight enumerator of a self-dual code over a ring of size s is in $\mathbb{C}[x^2 + (s-1)y^2, y(x-y)]$. See [27], Chap. 19 for a complete description modulo the fact that the MacWilliams relations for the Hamming weight enumerator for a code over a Frobenius ring of size s are identical to the classical MacWilliams relations.

For a binary Type II code, we can say more. The weight enumerator is also held invariant by the following matrix:

$$\begin{pmatrix} 1 & 0 \\ 0 & i \end{pmatrix},$$

where i is the complex square root of -1 since the Hamming weight of every vector is a multiple of 4.

The group G_{II} of matrices holding the weight enumerator invariant has order 192, and applying the classic theorem of Molien, we have that

$$\Phi(\lambda) = \frac{1}{(1-\lambda^8)(1-\lambda^{24})} = 1 + \lambda^8 + \lambda^{16} + 2\lambda^{24} + 2\lambda^{32} + \ldots \tag{5.6}$$

Table 5.1 The weight enumerator for a Type II [72, 36, 16] code

C_i	i
1	0, 72
249849	16, 56
18106704	20, 52
462962955	24, 48
4397342400	28, 44
16602715899	32, 40
25756721120	36

The generating invariants can be found easily by a straightforward computation which gives the well known Gleason's Theorem first proven in [24] by Gleason. The weight enumerator of a Type II self-dual code is a polynomial in $\mathbb{C}[x^8 + 14x^4y^4 + y^8, x^4y^4(x^4 - y^4)^4]$.

It is a direct consequence of this theorem that if C is a Type II $[n, k, d]$ code then

$$d \leq 4\lfloor \frac{n}{24} \rfloor + 4. \tag{5.7}$$

Any code that meets this bound is called an extremal code. Any code with parameters $[24k, 12k, 4k + 4]$ is such an extremal code. It is not known whether these codes exist until $24k \geq 3720$ at which a coefficient becomes negative. The existence of these codes has been a major question driving the study of binary self-dual codes.

For length 24, there is a [24, 12, 8] code, namely the well known extended Golay code. For length 48, the code also exists and is called the Pless code. See [31] for a description of both. Hence, the first unknown case is whether there exists a [72, 36, 16] code.

It first appeared in print as an open question in [33]. In [27] it was Research Problem 19.3. Despite being a celebrated problem since the 1970s, it remains an open question. A complete description of various approaches to the problem can be found in [18].

Question 5.2 Does there exist a Type II [72, 36, 16] code?

It is then easy to determine the weight enumerator for a putative [72, 36, 16] Type II code. It is given in Table 5.1.

It is well known that the existence of a Type II [72, 36, 16] code is equivalent to the existence of a Type I [70, 35, 14] code and that if these two codes exist, then 5-(72, 16, 78) designs exist.

The more general version of this question is the following.

Question 5.3 For which k does there exists a doubly-even self-dual binary $[24k, 12k, 4k + 4]$ code?

The reason we describe this problem here is that a great deal of work has been done in an attempt to find a code over a ring which may have a Gray image that would be the long sought after code. Up to this point, the technique has not solved the problem, but it has found many interesting binary codes.

One of the difficulties of this problem can be seen with respect to the very close relationship between the theory of self-dual codes and the theory of unimodular lattices as we have seen in Sect. 5.2. Because of this very close relationship there is often a parallel proof for similar results in both settings. For example, the proof of a bound on the minimum norm unimodular lattices often has a corresponding proof for a similar bound on the minimum weight of a code. Additionally, a non-existence proof of a lattice often has a corresponding proof for the non-existence of a code with related parameters. An important example of this is the work done by Conway and Sloane with respect to the shadow of codes and lattices, see [6]. It has been shown in [29] that an extremal lattice of length 72 exists. From the perspective of coding theory, this means that if the $[72, 36, 16]$ code does not exist, then the proof would have to be significantly different than the usual proofs in that it can have no corresponding proof for the extremal lattice. This fact eliminates many of the usual tools that coding theorists employ in similar situations.

Another reason for attempting to construct binary self-dual codes from codes over rings is that the classification of binary self-dual codes has been a productive and important area of research in coding theory for decades. See [5, 32] for early work in this vein and [13] for later work. In terms of finding binary self-dual codes from rings, see [19, 22, 23] for example.

Theorem 5.9 *Let C be a self-dual code over the ring R_k of length n and let ϕ_k be the Gray corresponding map. Then $\phi_k(C)$ is a binary self-dual code of length $2^k n$.*

Proof If $C = C^\perp$ we have $\phi_k(C) = \phi(C^\perp) = \phi_k(C)^\perp$ by Lemma 4.7. $\square$

Definition 5.2 A code C over a finite commutative Frobenius ring R is said to be formally self-dual if $W_C(x, y) = W_{C^\perp}(x, y)$.

Example 5.5 Let R be a finite commutative Frobenius ring. The code C generated by $(1, 0)$ has orthogonal generated by $(0, 1)$. Hence, C is a formally-self-dual code.

Since the weight enumerator of a formally self-dual code over a ring of cardinality s is held invariant by the action of the MacWilliams relations, it must be an element of $\mathbb{C}[x^2 + (s-1)y^2, y(x-y)]$.

Question 5.4 For any polynomial $p(x, y)$ of $\mathbb{C}[x^2 + (s-1)y^2, y(x-y)]$ with nonnegative coefficients, determine if there is a code C over a ring of size s with $W_C(x, y) = p(x, y)$.

In a more restricted setting, one of the most important questions has been the following.

Question 5.5 This question splits into two cases.

1. (Type I) For any polynomial $p(x, y)$ of $\mathbb{C}[x^2 + y^2, y(x - y)]$ with non-negative coefficients, determine if there is a binary code C with $W_C(x, y) = p(x, y)$.
2. (Type II) For any polynomial $p(x, y)$ of $\mathbb{C}[x^8 + 14x^4y^4 + y^8, x^4y^4(x^4 - y^4)^4]$ with non-negative coefficients, determine if there is a binary code C with $W_C(x, y) = p(x, y)$.

Notice that this important question is not only looking for codes which are formally self-dual (in terms of the definition we have given). It is possible for a code to be the non-linear image of a quaternary code and have a weight enumerator in the desired space. The heart of this question is trying to determine when putative optimal codes exist. For example, if a putative optimal code did not exist, but it was difficult to prove the non-existence it might be possible that a non-linear code with that weight enumerator existed. This would make any proof based solely on the weight enumerator of the putative code impossible. In this vein we have the following theorem.

Theorem 5.10 *Let C be a self-dual code over $\mathbb{Z}_4$. Then $\phi(C)$ has a weight enumerator fixed by the action of the MacWilliams relations.*

Proof Let C be a self-dual quaternary code. Then since $C = C^\perp$, we have $L_C(x, y) = L_{C^\perp}(x, y) = \frac{1}{|C|} L_C(x + y, x - y)$ by Theorem 4.1. Then since $L_C(x, y) = W_{\phi_4(C)}(x, y)$, we have the result. $\square$

As an example of this theorem consider the quaternary self-dual codes of length less than or equal to 8 as given in [7]. Their images are determined in [12] and are presented in Table 5.2. Notice that the image can be both linear or non-linear and self-dual or not self-dual, but all have weight enumerators that are fixed by the action of the MacWilliams relations.

Theorem 5.11 *Let C be a self-dual code over the ring S_k or T_k of length n and let ϕ_{k+1} be the corresponding Gray map. Then $\phi_{k+1}(C)$ is a binary code of length $2^{k+1}n$ which has a weight enumerator fixed by the action of the MacWilliams relations.*

Proof By Theorem 4.7, we have that $\psi_k(C)$ is a quaternary self-dual code. Then $\phi_{k+1}(C) = \phi(\psi_k(C))$, where ϕ is the Gray map defined for $\mathbb{Z}_4$. Since $\psi_k(C)$ is self-dual, we apply Theorem 5.10, and we have the result. $\square$

Table 5.2 Binary images of self-dual codes over $\mathbb{Z}_4$

Code	Length	Binary image	Orthogonality
$\mathcal{A}_1$	1	[2, 1, 2] Linear code	Self-dual
$\mathcal{D}_4^{\oplus}$	4	[8, 4, 4] Linear code	Self-dual
$\mathcal{D}_6^{\oplus}$	6	[12, 6, 4] Linear code	Not self-dual
$\mathcal{E}_7^{+}$	7	$(14, 2^7, 4)$ Non-linear code	Not self-dual
$\mathcal{D}_8^{\oplus}$	8	[16, 8, 4] Linear code	Not self-dual
$\mathcal{E}_8$	8	$(16, 2^8, 4)$ Non-linear code	Not self-dual
$\mathcal{K}_8$	8	[16, 8, 4] Linear code	Self-dual
$\mathcal{K}_8'$	8	[16, 8, 4] Linear code	Self-dual
$\mathcal{O}_8$	8	$(16, 2^8, 4)$ Non-linear code	Not self-dual
$\mathcal{Q}_8$	8	[16, 8, 4] Linear code	Not self-dual

5.4   Connections to Designs

Self-dual codes and designs have had a symbiotic relationship. Namely, self-dual codes have been used to construct designs and designs have been used to construct self-dual codes. The most powerful tool in constructing designs from codes is the well known Assmus-Mattson theorem which follows, see [2] for a complete description of this theorem.

Theorem 5.12 (Assmus-Mattson) *Let C be a code over $\mathbb{F}_q$ of length n with minimum weight d, and let $d^{\perp}$ denote the minimum weight of $C^{\perp}$. Let $w = n$ when $q = 2$ and otherwise the largest integer w satisfying $w - (\frac{w+q-2}{q-1}) < d$, define $w^{\perp}$ similarly. Suppose there is an integer t with $0 < t < d$ that satisfies the following condition: for $W_{C^{\perp}}(Z) = B_i Z^i$, at most $d - t$ of $B_1, B_2, \ldots, B_{n-t}$ are non-zero. Then for each i with $d \leq i \leq w$, the supports of the vectors of weight i of C, provided there are any, yield a t-design. Similarly, for each j with $d^{\perp} \leq j \leq min\{w^{\perp}, n - t\}$, the supports of the vectors of weight j in $C^{\perp}$, provided there are any, form a t-design.*

As an example of the power of this theorem, consider the binary [24, 12, 8] extended Golay code. Its weight enumerator is $W_C(1, y) = 1 + 759y^8 + 2576y^{12} + 759y^{16} + y^{24}$. By the above Assmus-Mattson theorem, the vectors of all weights hold 5-designs. In fact, it is easy to prove that any extremal [24k, 12k, 4k + 4] Type II binary code has the property that all of its non-trivial weights hold non-trivial 5-designs.

Note however that the theorem is stated for codes over a finite field. Some small generalizations for codes over rings have been made in this area, for example over $\mathbb{Z}_4$ in [34]. However, as of yet, no really interesting results have come by finding designs in the codes over rings. Hence, we state this as a question.

Question 5.6 Determine a technique for finding interesting designs using codes over rings.

There are well known constructions of self-dual codes over fields from designs; see [1] for a description. There have been some constructions of self-dual codes over rings, for example for $\mathbb{Z}_4$ in [21]. In general, these constructions required that there is a prime p that sharply divided the order of the design and then this prime was the characteristic of the underlying field. In [14], a construction was given, that did not require that there was a prime that sharply divided the order of the design, for self-dual codes over $\mathbb{Z}_m$. This was a generalization of the construction given in [11, 25]. In Glynn's construction the design was a finite projective plane and heavily used the geometry of that plane. The generalized construction in [14] did not require that the design was a plane, but rather a design from the broader class of symmetric designs. We shall now generalize this construction for finite commutative Frobenius rings.

We recall the definition of a design. A $t - (v, k, \lambda)$ design is a set of v points, with blocks of size k, such that any t points are incident with λ blocks. A symmetric design is a $2 - (v, k, \lambda)$ design where the number of points is equal to the number of lines and the points and blocks have identical incidence properties. In a symmetric design, the number of points incident with a block is $n + \lambda$ and the number of blocks incident with a point is $n + \lambda$.

Let L be a block. Through each point incident with the block there are $(n + \lambda - 1)$ blocks other than L. In this counting each block is counted exactly λ times. Therefore, in a symmetric design we have that the number of points is $v = \frac{(n+\lambda-1)(n+\lambda)}{\lambda} + 1$ and $k = n + \lambda$, where n is the order of the design.

When $\lambda = 1$, a symmetric design is a finite projective plane. When $\lambda = 2$, a symmetric design is a biplane.

Let $D = (\mathcal{P}, \mathcal{B}, \mathcal{I})$ be a (v, k, λ) symmetric design. If $\mathcal{B}' = \{b' \mid b'$ is the complement of a block in $\mathcal{B}\}$, then the complementary design $D^c = (\mathcal{P}, \mathcal{B}', \mathcal{I})$ is a $(v, v - k, b - 2r - \lambda)$, where b is the size of the blocks.

For this construction, we let D be a symmetric design of order n and we let m be an integer $n + 1$. We let R be a finite commutative Frobenius ring with characteristic m.

Denote the point set of D by $\mathcal{P}$ and the block set by $\mathcal{B}$. We denote the points by $\mathcal{P} = \{q_1, q_2, \ldots, q_{|\mathcal{P}|}\}$ and the blocks by $\mathcal{B} = \{\ell_1, \ell_2, \ldots, \ell_{|\mathcal{P}|}\}$. The ambient space for the codes we consider is $R^{\mathcal{P} \cup \mathcal{B}}$.

For a point $q \in \mathcal{P}$, we denote the characteristic function of the point by χ_q. That is, χ_q is the vector that is 1 at the coordinate corresponding to q and 0 elsewhere. Similarly, we let ψ_q be the vector that is 1 at the coordinate corresponding to ℓ, if q is incident with ℓ, and 0 elsewhere.

Define the following vector:

$$\Delta(q_i, q_j) = (\chi_i - \chi_j, \psi_i - \psi_j). \tag{5.8}$$

The weight of $\Delta(q_i, q_j)$ is $2n + 2$ where n is the order of the design.

The following lemma is a direct generalization of the lemma in [14].

Lemma 5.7 *Let D be a symmetric design of order n and let m be a positive integer dividing $n + 1$. Let R be a finite commutative Frobenius ring with characteristic m. Then we have that*

$$[\Delta(q_i, q_j), \Delta(q_{i'}, q_{j'})] = 0. \tag{5.9}$$

Proof We have three cases to consider.

Case 1: In this case, we assume that the points q_i are 4 distinct points. In this case, the supports of the vectors $\chi(q_i)$ are distinct. Then we have that

$$[(\psi_{q_i} - \psi_{q_j}), (\psi_{q_{i'}} - \psi_{q_{j'}})] = [\psi_{q_i}, \psi_{q_{i'}}] - [\psi_{q_i}, \psi_{q_{j'}}] - [\psi_{q_j}, \psi_{q_{i'}}] + [\psi_{q_j}, \psi_{q_{j'}}]$$
$$= \lambda - \lambda - \lambda + \lambda = 0.$$

Case 2: In this case, we assume that $q_i = q_{i'}$ and that $q_j \neq q_{j'}$. Then we have that

$$[(\chi_{q_i} - \chi_{q_j}), (\chi_{q_{i'}} - \chi_{q_{j'}})] = 1, \tag{5.10}$$

since the support of $\chi(q_i)$ is the support of $\chi(q_{i'})$ and the other supports are disjoint. This give that

$$[(\psi_{q_i} - \psi_{q_j}), (\psi_{q_{i'}} - \psi_{q_{j'}})] = [\psi_{q_i}, \psi_{q_i}] - [\psi_{q_i}, \psi_{q_{j'}}] - [\psi_{q_j}, \psi_{q_i}] + [\psi_{q_j}, \psi_{q_{j'}}]$$
$$= (n + \lambda) - \lambda - \lambda + \lambda = n.$$

Then we have that

$$[((\chi_{q_i} - \chi_{q_j}), (\psi_{q_i} - \psi_{q_j})), ((\chi_{q_{i'}} - \chi_{q_{j'}}), (\psi_{q_{i'}} - \psi_{q_{j'}}))] = 1 + n = 0. \tag{5.11}$$

Case 3: For the third case, we assume $q_i = q_{j'}$ and $q_j \neq q_{i'}$. In an argument similar to Case 2, we obtain

$$[((\chi_{q_i} - \chi_{q_j}), (\psi_{q_i} - \psi_{q_j})), ((\chi_{q_{i'}} - \chi_{q_{j'}}), (\psi_{q_{i'}} - \psi_{q_{j'}}))] = -(1 + n) = 0. \tag{5.12}$$
$$\square$$

Note that in this lemma nothing was assumed about the finite Frobenius ring except that it had a characteristic dividing $n + 1$.

We are now in a position to construct a self-orthogonal code. Define $C_R(D)$ to be

$$C_R(D) = \langle \Delta(q_i, q_j) \mid q_i, q_j \in \mathcal{P} \rangle. \tag{5.13}$$

We note that this code is self-orthogonal by Lemma 5.7.

From this code we can construct self-dual codes in a variety of ways depending on the structure of the ring R.

Define the matrix M_D to be the $|\mathcal{P}| - 1$ by $2|\mathcal{P}|$ matrix where the i-th row of M_D is $\Delta(q_1, q_{i+1})$. We have that the rows of M_D are mutually orthogonal over the ring R when the characteristic of the ring divides $n + 1$. By definition we have that M_D generates $C_R(D)$. Moreover, the structure of M_D gives that the code is a free code. It follows that the cardinality of $C_R(D)$ is $|R|^{|\mathcal{P}|-1}$.

We have proven the following.

Lemma 5.8 *Let D be a symmetric design of order n with m an integer dividing $n + 1$. Let R be a finite commutative ring with characteristic m where m divides $n + 1$. We have that $C_R(D)$ is a self-orthogonal, linear code with $|R|^{|\mathcal{P}|-1}$ elements.*

We define two additional vectors. Define P to be the vector that is 1 on the coordinates corresponding to the points and 0 on the coordinates corresponding to the blocks. Define L to be the vector that is 0 on the coordinates corresponding to the points and 1 on the coordinates corresponding to the blocks.

We have that both vectors are in the orthogonal of $C_R(D)$. Specifically, $P \in C_R(D)^\perp$ since $[\Delta(q_i, q_j), P] = 1 - 1 = 0$ and $L \in C_R(D)^\perp$ since $[\Delta(q_i, q_j), L] = n - n = 0$.

In order to make $\alpha P + \beta L$ a self-orthogonal vector, we need $[\alpha P + \beta L, \alpha P + \beta L] = 0$ which means that $(\alpha^2 + \beta^2)|\mathcal{P}| = 0$. If $|\mathcal{P}| \not\equiv 0 \pmod{m}$, this means that $\alpha^2 = -\beta^2$ so that the ring must have an element γ such that $\gamma^2 = -1$.

If the ring R has γ with $\gamma^2 = -1$ and m does not divide v, then we define the following code:

$$E_R(D) = \langle C_R(D), P + \gamma L \rangle. \tag{5.14}$$

We cannot use this description of $E_R(D)$ when m divides v and $\lambda - 1$ is the square root of -1 since then this the vector will already be in $C_R(D)$. We explain this now.

In a symmetric design, we have

$$\sum_{i=2}^{v} \Delta(q_1, q_i) = \sum_{i=2}^{v} ((\chi_{q_i} - \chi_{q_j}), (\psi_{q_i} - \psi_{q_j}))$$
$$= (v - 1, -1, -1, \ldots, -1, \alpha(1), \alpha(2), \ldots, \alpha(v)),$$

where

$$\alpha(i) = \begin{cases} -n - \lambda & \text{if } \ell_i \text{ is not incident with } q_1 \\ v - n - \lambda & \text{if } \ell_i \text{ is incident with } q_1. \end{cases}$$

Lemma 5.9 *Let D be a symmetric design of order n with m an integer dividing $n + 1$. Let R be a finite commutative ring with characteristic m where m divides $n + 1$. If m divides v, then $(-1, -1, \ldots, -1, -n - \lambda, -n - \lambda, \ldots, -n - \lambda) \in C_R(D)$.*

Proof Using the previous computation when m divides v, we have

$$\sum_{i=2}^{v} \Delta(q_1, q_i) = (-1, -1, \ldots, -1, -n - \lambda, -n - \lambda, \ldots, -n - \lambda). \qquad (5.15)$$

$\square$

Multiplying the above vector by -1 we have $(1, 1, \ldots, 1, n + \lambda, n + \lambda, \ldots, n + \lambda)$. Then we have $(n + \lambda)^2 = (\lambda - 1)^2$.

This leads to the following.

Lemma 5.10 *Let D be a symmetric design of order n with m an integer dividing $n + 1$. Let R be a finite commutative ring with characteristic m where m divides $n + 1$. If m divides v and $(\lambda - 1) = \sqrt{-1}$, then $P + \sqrt{-1}L \in C_R(D)$.*

In this case, when $P + \sqrt{-1}L \in C_R(D)$, we can define

$$E'_R(D) = \langle C_R(D), P + L \rangle. \qquad (5.16)$$

We know that $P + L$ is in $C_R(D)^{\perp}$. Then $[P + L, P + L] = 2v = 0$, and this gives the following.

Theorem 5.13 *Let D be a symmetric design of order n with m an integer dividing $n + 1$. Let R be a finite commutative ring of characteristic m. If m does not divide v or $(\lambda - 1)$ is not $\sqrt{-1}$ and $\sqrt{-1} \in \mathbb{Z}_m$, then $E_R(D)$ is a self-dual code over $\mathbb{Z}_m$ of length $2|\mathcal{P}|$. If m does divide v and $(\lambda - 1) = \sqrt{-1}$, then $E'_m(D)$ is a self-dual code over $\mathbb{Z}_m$ of length $2|\mathcal{P}|$.*

Proof The code is self-orthogonal by construction and its cardinality is $m|C_R(D)| = m^{|\mathcal{P}|}$. $\square$

Now we shall give a construction when the ring does not have a γ with $\gamma^2 = -1$, but m is a square where m is the characteristic of the ring. Set $m = q^2$ and define $F_R(D) = \langle C_R(D), qP, qL \rangle$.

Theorem 5.14 *Let R be a finite commutative Frobenius ring of characteristic m. Let D be a symmetric design of order n with $m = q^2$ an integer dividing $n + 1$. The code $F_R(D)$ is a self-dual code over R of length $2|\mathcal{P}|$.*

Proof It is immediate that $[qP, qP] = q^2 = 0$ and $[qL, qL] = q^2 = 0$. We have that $|F_R(D)| = q(q(|C_R(D)|)) = m|C_R(D)| = m^{|\mathcal{P}|}$, so $F_R(D)$ is a self-dual code. $\square$

5.5 Linear Complementary Dual

We shall now consider another family of codes, which in some sense are similar to self-dual codes in that they are described in terms of their relationship with their orthogonal. Specifically, it is the family of Linear Complementary Dual (LCD) codes.

Linear complementary dual codes were introduced by Massey in [28] and they give an optimum linear coding solution for the two user binary adder channel. In [4], these codes were used in counter measures to passive and active side channel analyses on embedded cryto-systems. Given the importance of self-dual codes and LCD codes, it seems natural to generalize these two families to codes whose intersection with its orthogonal has a given cardinality. We make such a generalization in this section.

Definition 5.3 A code C over a finite commutative Frobenius ring is a Linear Complementary Dual (LCD) code if $C \cap C^\perp = \{\mathbf{0}\}$.

Example 5.6 The trivial code R^n has dual $\{\mathbf{0}\}$ and hence is an LCD code. The code generated by $(1, 0)$ has an orthogonal generated by $(0, 1)$ and is an LCD code.

We can now generalize some results given first in [17].

Lemma 5.11 *Let* $\mathbf{v}_1, \mathbf{v}_2, \ldots, \mathbf{v}_k$ *be vectors over a finite commutative Frobenius ring such that* $[\mathbf{v}_i, \mathbf{v}_i] = 1$ *for each* i *and* $[\mathbf{v}_i, \mathbf{v}_j] = 0$ *for* $i \neq j$. *Then* $C = \langle \mathbf{v}_1, \mathbf{v}_2, \ldots, \mathbf{v}_k \rangle$ *is an LCD code over* R.

Proof A vector in the code C is of the form $\mathbf{w} = \sum \alpha_i \mathbf{v}_i$. If $\mathbf{w}$ is non-trivial, then there exists a j such that $\alpha_j \neq 0$. Then we have that $[\mathbf{v}_j, \mathbf{w}] = \alpha_j \neq 0$. This gives that $\mathbf{w} \notin C^\perp$. Therefore, we have that no non-trivial vector in C is also in $C^\perp$, which gives that $Hull(C) = \{\mathbf{0}\}$. $\qquad \square$

As an immediate consequence we have the following corollary.

Corollary 5.4 *Let* G *be a generator matrix for a code over a finite commutative Frobenius ring. Then* $GG^T = I_n$ *and* G *generates an LCD code.*

If the ring is a field then we can also say that if $det(GG^T) \neq 0$ then G generates an LCD code.

One way to construct these codes using designs is given in [17].

We recall that a Balanced Incomplete Block Design (BIBD) with parameters $t - (b, v, k, r, \lambda)$ is a set of v points, with blocks of size b, k points incident with each block, r blocks incident with every point, and through any t points there are λ blocks.

Theorem 5.15 *[17] Let* M *be the incidence matrix of a* $2 - (v, k, \lambda)$ *BIBD, where the columns are indexed by the points. If* $rk(r - \lambda) \neq 0 \pmod{p}$, *then* M *generates an LCD code over a field of characteristic* p.

Proof From Theorem 1.4.1 in [1], we have that

$$\det(MM^T) = rk(r - \lambda)^{v-1}.$$

Therefore, if $rk(r - \lambda) \neq 0 \pmod{p}$, we have that the row span of M is an LCD code.   □

We can also use the family of rings R_k, S_k, and T_k to construct LCD codes. Recall that ψ_k is their associated Gray map.

Theorem 5.16 *Let C be an LCD code over R_k. Then $\psi_k(C)$ is an LCD code over $\mathbb{F}_2$. Let C be an LCD code over S_k or T_k. Then $\psi_k(C)$ is an LCD code over $\mathbb{Z}_4$.*

Proof This follows from the fact that, in each case, $\psi_k(C^\perp) = \psi_k(C)^\perp$.   □

Given the importance of self-dual codes and LCD codes, it seems natural to generalize the idea of codes defined by their relationship to their duals. We begin with the following definition, which is inspired by the definition given in [2].

Definition 5.4 Let C be a code over a finite commutative Frobenius ring. The Hull of the code is defined to be $Hull(C) = C \cap C^\perp$.

Definition 5.5 Let C be a code over a finite field. Then we say that a code is an i-dual code if $Hull(C)$ has dimension i.

Then for a code over a finite field, we have that an $\frac{n}{2}$-dual code of length n is a self-dual code and a 0-dual code is an LCD code.

For codes over arbitrary rings we no longer have the notion of dimension; so we make the following generalized definition.

Definition 5.6 Let C be a code over a finite commutative Frobenius ring. Then we say that a code is an M-dual code if $|Hull(C)| = M$.

Consider a code of length n over a finite commutative ring R generated by the matrix $(I_k|A)$. This code has $|R|^k$ elements and is a free code. The code generated by $(-A^T|I_{n-k})$ has $|R|^{n-k}$ elements. The inner-product of the i-th row of $(I_k|A)$ and the j-th row of $(-A^T|I_{n-k})$ is $-A_{i,j} + A_{i,j} = 0$. These two facts give that the code $\langle(-A^T|I_{n-k})\rangle = \langle(I_k|A)\rangle^\perp$. From this result we can give the following theorem, which serves as an effective computational technique.

Theorem 5.17 *Let C be a code over a finite field of length n generated by $(I_k|A)$. Let D be the code generated by*

$$\begin{pmatrix} I_k & A \\ A^T & I_{n-k} \end{pmatrix}.$$

If $dim(D) = s$, then C is an $(n - s)$-dual code.

Proof In this scenario we have that $type(C) = \{k_0, 0, \ldots, 0\}$ and $type(C^{\perp}) = \{n - k_0, 0, \ldots, 0\}$. We have that $dim(C) + dim(C^{\perp}) - dim(Hull(C)) = n - dim(Hull(C))$. This gives that $dim(Hull(C)) = n - s$ and C is an $n - s$-dual code.   □

For a code over a finite field, this simplifies to the fact that if the generated code has dimension s, then C is an $n - 2$-dual code. As an immediate consequence, if the code generated by

$$\begin{pmatrix} I_k & A \\ A^T & I_{n-k} \end{pmatrix}.$$

has dimension n then the code is an LCD code and if it has dimension k with $k = \frac{n}{2}$ then the code is self-dual.

We can generalize Theorem 5.17 as follows.

Theorem 5.18 *Let C be a code over a finite commutative Frobenius ring of length n generated by $(I_k|A)$. Let D be the code generated by*

$$\begin{pmatrix} I_k & A \\ A^T & I_{n-k} \end{pmatrix}.$$

If $|\langle D \rangle| = S$, then C is a $\frac{|R|^n}{S}$-dual code.

Proof We have $|D| = \frac{|C||C^{\perp}|}{|Hull(C)|}$. This gives that $S = \frac{|R|^n}{|Hull(C)|}$ and C is a $\frac{|R|^n}{S}$-dual code.   □

We keep Theorems 5.17 and 5.18 separate since the first theorem is significantly easier to use. Namely, in Theorem 5.17, all that is required is to row reduce the given matrix. This can be done very easily. However, in Theorem 5.18, it is not always easy to determine the size of a code generated by a matrix over an arbitrary ring. Specifically, there is not always a general form that a generator matrix can be placed in for any code.

References

1. Assmus Jr., E.F., Key, J.D.: Designs and their Codes. Cambridge University Press, Cambridge (1992)
2. Assmus, E.F., Mattson, H.F.: New 5-designs. J. Comb. Theory **6**, 122–151 (1969)
3. Bannai, A., Dougherty, S.T., Harada, M., Oura, M.: Type II codes, even unimodular lattices, and invariant rings. IEEE-IT **45**(4), 1194–1205 (1999)
4. Carlet, C., Guilley, S.: Complementary dual codes for counter-measures to sidechannel attacks. In: Proceedings of the 4th ICMCTA Meeting, Palmela, Portugal (2014)
5. Conway, J.H., Pless, V., Sloane, N.J.A.: The binary self-dual codes of length up to 32: a revised enumeration. J. Combin. Theory Ser. A **60**, 183–195 (1992)
6. Conway, J.H., Sloane, N.J.A.: A new upper bound on the minimal distance of self-dual codes. IEEE Trans. Inform. Theory **IT-36**, 1319 - 1333 (1990)

7. Conway, J.H., Sloane, N.J.A.: Self-dual codes over the integers modulo 4. J. Combin. Theory Ser. A **62**(1), 30–45 (1993)
8. Conway, J. H., Sloane, N. J. A., Sphere packings, lattices and groups. Third edition. With additional contributions by E. Bannai, R. E. Borcherds, J. Leech, S. P. Norton, A. M. Odlyzko, R. A. Parker, L. Queen and B. B. Venkov. Grundlehren der Mathematischen Wissenschaften [Fundamental Principles of Mathematical Sciences], 290. Springer-Verlag, New York, (1999)
9. Choie, Y.J., Dougherty, S.T.: Codes over Σ_{2m} and Jacobi forms over the quaternions. Appl. Algebra. Eng. Commun. Comput. **15**(2), 129–147 (2004)
10. Choie, Y.J., Dougherty, S.T.: Codes over rings, complex lattices, and Hermitian modular forms. Eur. J. Comb. **26**(2), 145–165 (2005)
11. Dougherty, S.T.: A new construction of self-dual codes from projective planes. Australas. J. Comb. **31**, 337–348 (2005)
12. Dougherty, S.T., Fernandez-Cordoba, C.: Codes over $\mathbb{Z}_{2k}$, Gray maps and self-dual codes. Adv. Math. Commun. **54** (2011)
13. Dougherty, S.T., Harada, M., Gulliver, T.A.: Extremal binary self-dual codes, IEEE Trans. Inf. Theory, 2036–2047 (1997)
14. Dougherty, S.T., Gulliver, A., Ramadurai, R.: Symmetric designs and self-dual codes over rings. Ars Comb. **85**, 149–161 (2007)
15. Dougherty, S.T., Harada, M., Solé, P.: Self-dual codes over rings and the Chinese remainder theorem. Hokkaido Math. J. **28**, 253–283 (1999)
16. Dougherty, S.T., Kim, J.L., Kulosman, H., Liu, H.: Self-dual codes over Frobenius rings. Finite Fields Appl. **16**, 14–26 (2010)
17. Dougherty, S.T., Kim, J., Ozkaya, B., Sok, L., Solé, P.: The combinatorics of LCD codes: linear programming bound and orthogonal matrices, to appear in IJICOT
18. Dougherty, S.T., Kim, J.L., Solé, P.: Open problems in coding theory, Contemp. Math. **634**, 79–99 (2015)
19. Dougherty, S., Yildiz, B., Karadeniz, S.: Self-dual codes over R_k and binary self-dual codes. Eur. J. Pure Appl. Math. **6**(1), 89–106 (2013)
20. Harada, M., Miezaki, T.: An optimal odd unimodular lattice in dimension 72. Arch. Math. (Basel) **97**(6), 529–533 (2011)
21. Klemm, M.: Selbstduale code über dem ring der ganzen zahlen modulo 4. Arch. Math. (Basel) **53**, 201–207 (1989)
22. Karadeniz, S., Dougherty, S.T., Yildiz, B.: Constructing formally self-dual codes over R_k. Discrete Appl. Math. **167**, 188–196 (2014)
23. Karadeniz, S., Yildiz, B.: New extremal binary self-dual codes of length 64 from R_3-lifts of the extended binary Hamming code. Des. Codes Cryptogr. **74**(3), 673–680 (2015)
24. Gleason, A.M.: Weight polynomials of self-dual codes and the MacWilliams identities. Actes Congres Internl. Math. **3**, 211–215 (1970)
25. Glynn, D.: The construction of self-dual binary codes from projective planes of odd order. Australas. J. Comb. **4**, 277–284 (1991)
26. Lam, C.W.H.: The search for a finite projective plane of order 10. Amer. Math. Monthly **98**(4), 305–318 (1991)
27. MacWilliams, F.J., Sloane, N.J.A.: The Theory of Error-Correcting Codes. North-Holland, Amsterdam (1977)
28. Massey, J.L.: Linear codes with complementary duals. Discrete Math. **106–107**, 337–342 (1992)
29. Nebe, G.: An even unimodular 72-dimensional lattice of minimum 8. J. Reine Angew. Math. **673**, 237–247 (2012)
30. Nebe, G., Rains, E. M., Sloane, N. J. A.: Self-dual codes and invariant theory. Algorithms Comput. Math. **17**. Springer, Heidelberg (2006)
31. Rains, E. M., Sloane, N. J. A.: Self-Dual Codes. Handbook of Coding Theory, vol. I, II, pp. 177–294. North-Holland, Amsterdam (1998)

32. Pless, V., Sloane, N.J.A.: On the classification and enumeration of self-dual codes. J. Combin. Theory Ser. A **18**, 313–335 (1975)
33. Sloane, N. J. A.: Is there a $(72, 36)$, $d = 16$ self-dual code? IEEE Trans. Inf. Theory IT **192**, 251 (1973)
34. Tanabe, K.: An Assmus-Mattson theorem for $\mathbb{Z}_4$-codes. IEEE Trans. Inf. Theory **46**(1), 48–53 (2000)

Chapter 6
Cyclic and Constacyclic Codes

Cyclic codes are one of the most widely studied families of codes, both because of their use in applications, and because of their rich algebraic structure. They were first introduced by Prange in [25]. Cyclic codes have also been generalized in numerous ways, including polycyclic, negacyclic, constacyclic, quasicyclic and skew cyclic codes. In this chapter, we shall attempt to take a very general view of these families of codes and couch them in an algebraic setting.

6.1 Polycyclic Codes

We begin with a very general algebraic description of a large family of codes. Let R be a finite commutative Frobenius ring. If $\mathbf{v} = (c_0, c_1, \ldots, c_{n-1})$ is a vector in R^n, then there is a natural connection to polynomials in $R[x]$ by viewing the vector as the polynomial $\sum_{i=0}^{n-1} c_i x^i$. This allows us to associate codes, in the traditional sense, with ideals in a polynomial ring. We begin with the standard definition of the various types of codes in this family.

Definition 6.1 Let $f(x)$ be a polynomial in $R[x]$, where R is a finite commutative Frobenius ring. A polycyclic code C over R is an ideal in $R[x]/\langle f(x) \rangle$.

- If $f(x) = x^n - 1$, then the code is said to be a cyclic code.
- If $f(x) = x^n + 1$, then the code is said to be negacyclic.
- If $f(x) = x^n + \lambda$, λ a unit, the code is said to be constacyclic.

The following conditions on the codewords follow immediately from the definition of cyclic, negacyclic, and constacyclic. A cyclic code satisfies the following:

$$(c_0, c_1, \ldots, c_{n-1}) \in C \Rightarrow (c_{n-1}, c_0, c_1, \ldots, c_{n-2}) \in C.$$

© The Author(s) 2017
S.T. Dougherty, *Algebraic Coding Theory Over Finite Commutative Rings*,
SpringerBriefs in Mathematics, DOI 10.1007/978-3-319-59806-2_6

A negacyclic code satisfies the following:

$$(c_0, c_1, \dots, c_{n-1}) \in C \Rightarrow (-c_{n-1}, c_0, c_1, \dots, c_{n-2}) \in C.$$

A constacyclic code satisfies the following:

$$(c_0, c_1, \dots, c_{n-1}) \in C \Rightarrow (\lambda c_{n-1}, c_0, c_1, \dots, c_{n-2}) \in C.$$

For much of this section, we prove results for constacyclic codes since the result then applies to negacyclic and cyclic by letting $\lambda = -1$ and $\lambda = 1$.

Example 6.1 The perfect binary Golay code is a cyclic code with parameters [23, 12, 7] and is generated by the polynomial $1 + x^2 + x^4 + x^5 + x^6 + x^{10} + x^{11}$. The perfect ternary Golay code is a cyclic code with parameters [11, 6, 5] and is generated by the polynomial $x^5 + x^4 - x^3 + x^2 - 1$.

The major drawback of a polycyclic code in general is that the orthogonal may not be a polycyclic code. In [11], it was shown that this can be remedied by defining an alternate duality for codes over finite fields. However, we can prove that for a constacyclic code the orthogonal is again a constacyclic code. We first introduce a function. For any vector $\mathbf{c} = (c_0, c_1, \dots, c_{n-1})$, let $\tau_\lambda(\mathbf{c}) = (\lambda c_{n-1}, c_0, c_1, \dots, c_{n-2})$.

Theorem 6.1 *Let R be a finite commutative Frobenius ring. If C is a constacyclic code over R of length n, then $C^\perp$ is a constacyclic code of length n over R.*

Proof Let $\mathbf{c}$ be any vector in C, where C is a λ-constcyclic code. Let $\mathbf{d} \in C^\perp$. Then $[\mathbf{d}, \tau_\lambda^i(\mathbf{c})] = 0$ for $i = 0$ to $i = n - 1$. It is a simple computation to see that this is equivalent to the statement $[\mathbf{c}, \tau_{\lambda^{-1}}^i(\mathbf{d})] = 0$ for $i = 0$ to $i = n - 1$. Therefore $C^\perp$ is a λ^{-1}-constacyclic code. □

Given this theorem, it is apparent why λ is always chosen to be a unit.

Corollary 6.1 *Let R be a finite commutative Frobenius ring. If C is a cyclic code over R of length n, then $C^\perp$ is a cyclic code. If C is a negacyclic code over R of length n, then $C^\perp$ is a negacyclic code.*

Proof The result follows immediately from Theorem 6.1, noting that both 1 and -1 are their own multiplicative inverses and hence they are units in any ring. □

We can use the Chinese Remainder Theorem, in the usual way, to show that the important object to study is constacyclic codes over local rings.

Theorem 6.2 *Let R be a finite commutative Frobenius ring and let $R = CRT(R_1, R_2, \dots, R_s)$ be the decomposition of R via the Chinese Remainder Theorem. If C_i is a λ_i-constacyclic code of length n over each R_i, where and $\lambda = CRT(\lambda_1, \lambda_2, \dots, \lambda_2)$, then $C = CRT(C_1, C_2, \dots, C_s)$ is a λ-constacyclic code over R.*

Proof Let $\mathbf{c}^i = (c_0^i, c_1^i, \ldots, c_{n-1}^i)$ be an arbitrary element in C_i. Then we have that $\tau(\mathbf{c}^i) = (\lambda_i c_{n-1}^i, c_0^i, \ldots, c_{n-2}^i)$ is an element of C_i, since it is λ_i-constacyclic, where τ represents the constacyclic action. Then any codeword $\mathbf{c} = (c_0, c_1, \ldots, c_{n-1})$ of C is of the form $\mathbf{c} = CRT(\mathbf{c}^1, \mathbf{c}^2, \ldots, \mathbf{c}^s)$. Then $CRT(\tau(\mathbf{c}^1), \tau(\mathbf{c}^2), \ldots, \tau(\mathbf{c}^s)) \in C$ and $CRT(\tau(\mathbf{c}^1), \tau(\mathbf{c}^2), \ldots, \tau(\mathbf{c}^s)) = (\lambda c_{n-1}, c_1, \ldots, c_{n-2})$. Therefore, C is a λ-constacyclic code. $\qquad\square$

It is clear that since we are looking for ideals in $R[x]/\langle x^n - \lambda \rangle$, we are in general looking for divisors of the polynomial $x^n - \lambda$ in $R[x]$. We describe a few examples.

Example 6.2 If the alphabet is a finite field, then every cyclic code C is generated by a nonzero monic polynomial of minimal degree in C, which must be a divisor of $X^n - 1$ by the minimality of degree. The ring $\mathbb{F}_q[x]$ is a unique factorization domain and so the factorization of $x^n - 1$ determines all cyclic codes. If $gcd(n, p) = 1$, then $x^n - \lambda$ has no repeated roots and $\mathbb{F}_q[x]/\langle x^n - \lambda \rangle$ is a semi-simple ring. If the characteristic of the field and the length are not relatively prime, then we are in the repeated root case.

Cyclic codes over $\mathbb{Z}_{p^e}$ were first studied by Calderbank and Sloane in [8]. Following this, Kanwar and López-Permouth provided a different approach in [20].

Example 6.3 Let n be the length of a code over the ring $\mathbb{Z}_{p^e}$ and assume n and p are relatively prime. Then $x^n - 1$ factors uniquely over $\mathbb{Z}_{p^e}$ by Hensel's Lemma. Cyclic codes over $\mathbb{Z}_{p^e}$ have the form $\langle f_0, pf_1, p^2 f_2, \ldots, p^{e-1} f_{e-1} \rangle$, where $f_{e-1} \mid f_{e-2} \mid \cdots \mid f_0 \mid x^n - 1$. These ideals are, in fact, principal and can be described as follows:

$$\langle f_0, pf_1, p^2 f_2, \ldots, p^{e-1} f_{e-1} \rangle = \langle f_0 + pf_1 + p^2 f_2 + \cdots + p^{e-1} f_{e-1} \rangle. \qquad (6.1)$$

See [8] for a complete description.

Notice in the previous example, that this case is very similar to the case for finite fields, in that the unique factorization of $x^n - 1$ easily determines all cyclic codes when the length is relatively prime to the characteristic of the field.

Recall that a code is cyclic over a local Frobenius non-chain ring of order 16 if and only if it is an ideal in the ring $\mathfrak{R}_n = R[x]/\langle x^n - 1 \rangle$, see [10]. The next theorem describes the ideals in $\mathfrak{R}_n$.

Example 6.4 If R is a commutative local ring of order 16 then the maximal ideal is generated by two elements u and v. Let n be an odd integer and $x^n - 1 = f_1(x) f_2(x) \cdots f_r(x)$ be the representation of $x^n - 1$ as a product of basic irreducible pairwise coprime polynomials in $R[x]$. Let $\hat{f}_i(x)$ denote the product of all $f_j(x)$ except $f_i(x)$. Then any ideal in $R[x]/\langle x^n - 1 \rangle$ is a sum of the following ideals, $\langle \hat{f}_i(x) \rangle$, $\langle u \hat{f}_i(x), v \hat{f}_i(x) \rangle$ and $\langle \alpha \hat{f}_i(x) \rangle$, $\alpha \in \{u, v, u + v, w\}$. See [10] for a complete description of cyclic codes over these rings.

Example 6.5 In [24], it was shown that a cyclic code of odd length over $\mathbb{Z}_4$ is of the form $C = \langle fh, 2fg \rangle$, where $fgh = x^n - 1$. Let $\mathcal{R}(C)$ be the quaternary code that is the preimage of the binary code generated by the image under the Gray map. Let $\mathcal{K}(C)$ be the preimage of the binary kernel. Then $\mathcal{R}(C)$ and $\mathcal{K}(C)$ are both cyclic codes. See [9], for a complete description.

These examples lead us to the natural fundamental question of constacyclic codes.

Question 6.1 Let R be a finite commutative Frobenius ring. Find all ideals in $R[x]/\langle x^n - \lambda \rangle$, where λ is a unit in R.

This question really has two parts. The first is determining the structure of ideals in the polynomial ring. The second is actually finding the specific ideals for a given length in a computational manner. For example, one can find all cyclic codes over a field by factoring $x^n - 1$ over that field. However, this leads to the question of factoring $x^n - 1$ for all n. In a ring, the question is more complicated due to the existence of zero divisors. For example, the code $\langle (2) \rangle$ is a cyclic code of length 1 over $\mathbb{Z}_4$ but has nothing to do with the factorization of $x - 1$ over $\mathbb{Z}_4$.

This question is usually divided into two distinct parts. The first, which is significantly easier, is when the length n is relatively prime to the characteristic of the ring. The second, which is often more difficult, is the so called repeated root case, when n is not relatively prime to the characteristic of the ring.

6.2 Constacyclic Codes Over Formal Power Series Rings and Chain Rings

In [27], Wan gave the structure of cyclic codes over Galois rings. This was extended in [23] by Norton and Sălăgean to finite chain rings. Dinh and López-Permouth generalized the structure of cyclic codes to finite chain rings in a more general setting in [7]. Following these papers, Dougherty and Park studied general properties of cyclic codes of length n over $\mathbb{Z}_m$, where n is an arbitrary integer, that is, not necessarily relatively prime to the characteristic of the ring.

Given the result in Theorem 6.2, which says that any λ-constacyclic code over a principal ideal ring is the product under the Chinese Remainder Theorem of constatcyclic codes over chain rings, it suffices to study codes over chain rings to understand λ-constacyclic codes over principal ideal rings. To study λ-constacyclic codes over chain rings, we shall take a very general view and study codes over formal power series rings.

Calderbank and Sloane first started this view of cyclic codes in [8] by studying codes over the ring of p-adic integers. They gave the structure of cyclic codes of length n over the p-adic integers, in the case when $\gcd(n, p) = 1$. Following this, Dougherty, Kim and Park extended this work by studying the lifts of codes over $\mathbb{Z}_p$ to $\mathbb{Z}_{p^e}$ and to the p-adics. In [16], Dougherty, Liu and Park defined a series of finite chain rings and introduced the concept of γ-adic codes over a formal power series ring. In [14], cyclic codes over these rings were studied. We shall describe the results given there in this section.

We begin by defining a family of rings.

Let $\mathbb{F}_q$ be an arbitrary finite field. Let i be an arbitrary positive integer. Define the following rings:

$$\mathcal{A}_i = \{a_0 + a_1\gamma + \cdots + a_{i-1}\gamma^{i-1} \mid a_i \in \mathbb{F}_q\} \tag{6.2}$$

where $\gamma^{i-1} \neq 0$, but $\gamma^i = 0$ in $\mathcal{A}_i$. The operations in this ring are defined as follows:

$$\sum_{l=0}^{i-1} a_l\gamma^l + \sum_{l=0}^{i-1} b_l\gamma^l = \sum_{l=0}^{i-1}(a_l + b_l)\gamma^l, \tag{6.3}$$

$$\sum_{l=0}^{i-1} a_l\gamma^l \cdot \sum_{l'=0}^{i-1} b_{l'}\gamma^{l'} = \sum_{s=0}^{i-1}(\sum_{l+l'=s} a_lb_{l'})\gamma^s. \tag{6.4}$$

We define the formal power series ring $\mathcal{A}_\infty$ as follows:

$$\mathcal{A}_\infty = \mathbb{F}_q[[\gamma]] = \{\sum_{l=0}^{\infty} a_l\gamma^l \mid a_l \in \mathbb{F}_q\}. \tag{6.5}$$

For each $i < \infty$, the ring $\mathcal{A}_i$ is a chain ring with maximal ideal $\langle\gamma\rangle$. In each case, $\mathcal{A}_i/\langle\gamma\rangle \cong \mathbb{F}_q$. The group of units of the infinite ring is $R_\infty^\times = \{\sum_{j=0}^{\infty} a_j\gamma^j \mid a_0 \neq 0\}$. Note that a_0 is an element in a field. This implies that if it is not zero, then it is necessarily a unit. The infinite ring $\mathcal{A}_\infty$ is a principal ideal domain.

We shall define a family of maps which serve to project from a larger ring to a smaller ring. Let i, j be two non-negative integers with $i \leq j$ or let i be a non-negative integer and let j be ∞. Then

$$\Psi_i^j : \mathcal{A}_j \to \mathcal{A}_i, \tag{6.6}$$

$$\sum_{l=0}^{j-1} a_l\gamma^l \mapsto \sum_{l=0}^{i-1} a_l\gamma^l. \tag{6.7}$$

It is immediate that each map of the form Ψ_i^j is a ring homomorphism.

We have seen that codes over a chain ring have an easily described generator matrix, see Theorem 2.12. Let C be a nonzero linear code over $\mathcal{A}_\infty$ of length n, it is shown in [14] that any generator matrix of C is permutation equivalent to a matrix of the following form:

$$G = \begin{pmatrix} \gamma^{m_0} I_{k_0} & \gamma^{m_0} A_{0,1} & \gamma^{m_0} A_{0,2} & \gamma^{m_0} A_{0,3} & & & \gamma^{m_0} A_{0,r} \\ & \gamma^{m_1} I_{k_1} & \gamma^{m_1} A_{1,2} & \gamma^{m_1} A_{1,3} & & & \gamma^{m_1} A_{1,r} \\ & & \gamma^{m_2} I_{k_2} & \gamma^{m_2} A_{2,3} & & & \gamma^{m_2} A_{2,r} \\ & & & \ddots & \ddots & & \\ & & & & \ddots & \ddots & \\ & & & & & \gamma^{m_{r-1}} I_{k_{r-1}} & \gamma^{m_{r-1}} A_{r-1,r} \end{pmatrix}, \quad (6.8)$$

where $0 \leq m_0 < m_1 < \cdots < m_{r-1}$ for some integer r.

The column blocks have sizes $k_0, k_1, \cdots, k_r$ and the k_i sum to n. With this generator matrix, we say that the code has type

$$(\gamma^{m_0})^{k_0} (\gamma^{m_1})^{k_1} \cdots (\gamma^{m_{r-1}})^{k_{r-1}}. \quad (6.9)$$

It is immediate that $k = k_0 + k_1 + \cdots + k_{r-1}$ is the rank of the code as a module. We refer to such codes as γ-adic codes.

In general, for linear γ-adic codes, we have that $C \subseteq (C^\perp)^\perp$. Unlike, for codes over finite rings, we do not always have $C = (C^\perp)^\perp$. As an example, let C be a code of length 3 over $\mathcal{A}_\infty$ generated by (γ, γ, γ). We have that $C^\perp = \{(a, b, c) \mid a + b + c = 0\}$. Then we have that $(C^\perp)^\perp$ contains the vector $(1, 1, 1)$. However, this vector is not in C. Therefore, $C \neq (C^\perp)^\perp$. The reason that this can occur is that the infinite ring $\mathcal{A}_\infty$ contains no zero-divisors, even though it is not a field. Therefore, if $\mathbf{c} \in C^\perp$ and $\mathbf{c} = \alpha \mathbf{v}$, then $[\mathbf{c}, \mathbf{w}] = 0$ for all $\mathbf{w} \in C$, which gives $[\alpha \mathbf{v}, \mathbf{w}] = \alpha[\mathbf{v}, \mathbf{w}] = 0$. This implies that $[\mathbf{v}, \mathbf{w}] = 0$, which gives that the code $C^\perp$ must have type 1^m for some m.

We say that a linear code C over $\mathcal{A}_\infty$ is basic if $C = (C^\perp)^\perp$. This definition was first given in [16].

Extend the map Ψ_i^j in the natural way to $\mathcal{A}_j[x]$, namely Ψ_i^j acts on the coefficients of the polynomials and fixes the indeterminate x. For a polynomial $f(x) \in \mathcal{A}_\infty[x]$, if $\deg(f(x)) > 0$ and $\gcd(a_0, a_1, \cdots, a_n) = 1$, then we say that $f(x)$ a primitive polynomial.

We shall prove some technical results to study constacyclic codes over $\mathcal{A}_\infty$. We begin with a lemma that characterizes primitive polynomials in terms of the projections. It first appears in [16].

Lemma 6.1 *Let $f(x)$ be a polynomial in $\mathcal{A}_\infty[x]$ with $\deg(f(x)) > 0$. Then $f(x)$ is a primitive polynomial if and only if $\Psi_i^\infty(f(x)) \neq 0$ for all $i < \infty$.*

Proof Assume that the polynomial is primitive and that there exists i such that $\Psi_i^\infty(f(x)) = 0$. Then all nonzero coordinates a_j of $f(x)$ must have the form $a_j = \gamma^{l_j} b_j$ with $l_j \geq i$. It follows that $\gcd(a_0, a_1, \cdots, a_n) = \gamma^m$ for some $m \geq i$. This contradicts that the polynomial is primitive.

Conversely, assume that $f(x)$ is not a primitive polynomial over $\mathcal{A}_\infty$. This implies that $\gcd(a_0, a_1, \cdots, a_n) = \gamma^i$ for some i. It follows that $\Psi_i^\infty(f(x)) = 0$. Then we have the result. $\square$

We relate polynomials to primitive polynomials in the following theorem.

Theorem 6.3 *Let $f(x)$ be a polynomial in $\mathcal{A}_\infty[x]$ such that $\deg(f(x)) > 0$. Then there exist a unique s and a primitive polynomial $g(x)$ such that $f(x) = \gamma^s g(x)$.*

Proof Let $f(x) = a_0 + a_1 x + \cdots + a_n x^n \in \mathcal{A}_\infty[x]$. Any nonzero element a_j of the ring can be written as $a_j = \gamma^j b_j$, where $j \geq 0$, and the element b_j is a unit. Set $s = \min\{l_j \mid 0 \neq a_j = \gamma^{l_j} b_j\}$. Then, we have

$$f(x) = \gamma^s(\gamma^{l_0-s} b_0 + \gamma^{l_1-s} b_1 x + \cdots + b_s x^s + \cdots + \gamma^{l_n-s} b_n x^n). \qquad (6.10)$$

Let

$$g(x) = \gamma^{l_0-s} b_0 + \gamma^{l_1-s} b_1 x + \cdots + b_s x^s + \cdots + \gamma^{l_n-s} b_n x^n. \qquad (6.11)$$

This gives that $f(x) = \gamma^s g(x)$.

The greatest common divisor $\gcd(\gamma^{l_0-s} b_0, \gamma^{l_1-s} b_1, \cdots, b_s, \cdots, \gamma^{l_n-s} b_n) = 1$ since $b_s \in R_\infty^\times$. It follows that $g(x)$ is a primitive polynomial. $\qquad \square$

We say that two polynomials $f(x)$ and $g(x) \in \mathcal{A}_i$ are coprime if there exist $u(x), v(x) \in \mathcal{A}_i[x]$ such that $f(x)u(x) + g(x)v(x) = 1$. Coprime polynomials $f(x)$ and $g(x)$ satisfy $\langle f(x) \rangle + \langle g(x) \rangle = \mathcal{A}_i[x]$.

If a polynomial $f(x)$ is reducible over $\mathcal{A}_\infty$ then there exist polynomials $g(x), h(x)$ such that $f(x) = g(x)h(x)$ and $0 < \deg(g(x)), \deg(h(x)) < \deg(f(x))$. This implies that

$$\Psi_i^\infty(f(x)) = \Psi_i^\infty(g(x)h(x)) = \Psi_i^\infty(g(x))\Psi_i^\infty(h(x)). \qquad (6.12)$$

Assuming $f(x)$ is a monic polynomial, we have that

$$0 < \deg(\Psi_i^\infty(g(x))), \deg(\Psi_i^\infty(h(x))) < \deg(\Psi_i^\infty(f(x))) = \deg(f(x)). \qquad (6.13)$$

It follows that if $f(x)$ is a monic polynomial in $\mathcal{A}_\infty[x]$, and $\Psi_i^\infty(f(x))$ is irreducible in $\mathcal{A}_i[x]$ for some $i < \infty$, then $f(x)$ must be irreducible over $\mathcal{A}_\infty$.

We can now study constacyclic codes over $\mathcal{A}_\infty$. We assume that the characteristic of the finite field is p and that the length of the code n is relatively prime to p.

Recall that for constacyclic codes we require λ to be a unit. Therefore, let λ be an arbitrary unit of $\mathcal{A}_\infty$. Define the map P_λ in the usual manner. Namely,

$$P_\lambda : R_\infty^n \to \mathcal{A}_\infty[x]/\langle x^n - \lambda \rangle,$$
$$P_\lambda(a_0, a_1, \cdots, a_{n-1}) = a_0 + a_1 x + \cdots + a_{n-1} x^{n-1} + \langle x^n - \lambda \rangle.$$

We note that a linear code $\mathcal{C}$ of length n over $\mathcal{A}_\infty$ is a λ-cyclic code if and only if $P_\lambda(\mathcal{C})$ is an ideal of $\mathcal{A}_\infty[x]/\langle x^n - \lambda \rangle$.

We extend the map Ψ_i^∞ in the canonical way. That is,

$$\Psi_i^\infty : \mathcal{A}_\infty[x]/\langle x^n - \lambda \rangle \to \mathcal{A}_i[x]/\langle x^n - \lambda \rangle. \qquad (6.14)$$

This map is a homomorphism. Therefore, if I is an ideal of $\mathcal{A}_\infty[x]/\langle x^n - \lambda \rangle$, then $\Psi_i^\infty(I)$ is an ideal of $\mathcal{A}_i[x]/\langle x^n - \lambda \rangle$. It follows that $\Psi_i^\infty P_1 = P_1 \Psi_i^\infty$.

To show that the image of a constacyclic code is again constacyclic under the map Ψ_i^∞ we need a technical lemma.

Lemma 6.2 *An element $\lambda \in \mathcal{A}_\infty$ is a unit if and only if $\Psi_i^\infty(\lambda)$ is a unit for all i.*

Proof An element in $\mathcal{A}_\infty$ is a unit if and only if a_0 is non-zero, where a_0 is the coefficient of γ^0. By noticing that the map Ψ_i^∞ fixes the coefficient of γ^0 we have the result. □

Theorem 6.4 *If C is a λ-constacyclic code over $\mathcal{A}_\infty$, then $\Psi_i^\infty(C)$ is a $\Psi_i^\infty(\lambda)$-constacyclic code over $\mathcal{A}_i$, for all $i < \infty$.*

Proof First we note that by Lemma 6.2, we have that $\Psi_i^\infty(\lambda)$ is a unit for all i. Let C be a constacyclic code over $\mathcal{A}_\infty$, then $P_1(C)$ is an ideal of $\mathcal{A}_\infty[x]/\langle x^n - 1 \rangle$. By the homomorphism in Eq. (6.14), we have that $\Psi_i^\infty(P_1(C)) = P_1(\Psi_i^\infty(C))$ is an ideal of $\mathcal{A}_i[x]/\langle x^n - 1 \rangle$. This gives the result. □

We have seen that the orthogonal of a constacyclic code over a finite ring is constacyclic. For the ring, $\mathcal{A}_\infty$ the proof is identical. Moreover, we have the following theorem.

Theorem 6.5 *Let C be a constacyclic code over $\mathcal{A}_\infty$. Then the code $\Psi_i^\infty(C^\perp)$ is a constacyclic code. If C is basic then $\Psi_i^\infty(C^\perp) = \Psi_i^\infty(C)^\perp$, for all $i < \infty$.*

Proof By Theorem 6.4, we have that $\Psi_i^\infty(C^\perp)$ is a constcyclic code for all $i < \infty$ since $C^\perp$ is a constacyclic code.

Let $\mathbf{v} \in \Psi_i^\infty(C^\perp)$ and let $\mathbf{w}$ be an element of $\Psi_i^\infty(C)$. This implies that there exist $\mathbf{v}' \in C^\perp$ and $\mathbf{w}' \in C$ such that $\mathbf{v} = \Psi_i^\infty(\mathbf{v}')$ and $\mathbf{w} = \Psi_i^\infty(\mathbf{w}')$. It follows that

$$[v, w] = [\Psi_i^\infty(v'), \Psi_i^\infty(w')] = \Psi_i^\infty[v', w'] = \Psi_i^\infty(0) = 0. \qquad (6.15)$$

Therefore, we have that $\Psi_i^\infty(C^\perp) \subseteq (\Psi_i^\infty(C))^\perp$.

We know that $C^\perp$ has type 1^{n-k} since it is the orthogonal of a code over $\mathcal{A}_\infty$. If C is basic, we have that $C = (C^\perp)^\perp$. This implies that C has type 1^k. Therefore, $\Psi_i^\infty(C^\perp)$ has type 1^{n-k} and $(\Psi_i^\infty(C))^\perp$ has type 1^{n-k}. This gives $(\Psi_i^\infty(C))^\perp = \Psi_i^\infty(C^\perp)$. □

Let C be a linear non-basic constacyclic code over $\mathcal{A}_\infty$, then the code $C' = (C^\perp)^\perp$ has type 1^m for some m and $C \subseteq C'$. This gives that if C is a linear non-basic constacyclic code over $\mathcal{A}_\infty$, then we can find a linear basic constacyclic code C' with $C \subseteq C'$. As a simple example, consider the code $C = \langle (\gamma, \gamma, \ldots, \gamma) \rangle$. This code is cyclic (1-constacyclic) and $(C^\perp)^\perp = \langle (1, 1, \ldots, 1) \rangle$ which is cyclic and contains C.

We need a result which allows us to lift the factorization of a polynomial. Specifically, we need Hensel's lemma whose proof can be found in [21].

Lemma 6.3 (Hensel's Lemma) *Let $f(x)$ be a polynomial over $\mathcal{A}_i$, where $i < \infty$, assume $\Psi_1^i(f(x)) = g_1(x)g_2(x)\cdots g_r(x)$ where $g_1(x), g_2(x),\cdots, g_r(x)$ are pairwise coprime polynomials over $\mathbb{F}_q$. Then there exist pairwise coprime polynomials $f_1(x), f_2(x),\cdots, f_r(x)$ over $\mathcal{A}_i$ such that*

$$f(x) = f_1(x)f_2(x)\cdots f_r(x)$$

and $\Psi_1^i(f_j(x)) = g_j(x)$, for $j = 1, 2, \cdots, r$.

Recall that if $f(x)$ is a monic irreducible divisor of $x^n - 1$ over $\mathbb{F}_q$, then $\langle f(x) \rangle$ is a prime ideal in $\mathbb{F}_q[x]/\langle x^n - 1 \rangle$.

Lemma 6.4 *Let i be a positive integer and let $\mathfrak{a}$ be a prime ideal in $\mathcal{A}_i[x]/\langle x^n - \lambda \rangle$, where λ is a unit in $\mathcal{A}_i$. Then $\gamma \in \mathfrak{a}$.*

Proof We know that $\gamma^i = 0 \in \mathfrak{a}$, since the nilpotency index of γ is i. It follows that either γ^{i-1} or γ is in $\mathfrak{a}$. If $\gamma \in \mathfrak{a}$ we are done. Otherwise, $\gamma^{i-1} \in \mathfrak{a}$. Then proceeding by induction, we have the result.   □

Theorem 6.6 *Let i be a positive integer and let λ be a unit in $\mathcal{A}_i$. The prime ideals in $\mathcal{A}_i[x]/\langle x^n - \lambda \rangle$ are of the form $\langle f_i(x), \gamma \rangle$, where $f_i(x)$ is any monic irreducible divisor of $x^n - \lambda$ over $\mathcal{A}_i$.*

Proof Let $\mathfrak{a}$ be a prime ideal in $\mathcal{A}_i[x]/\langle x^n - \lambda \rangle$. It follows that $\Psi_1^i(\mathfrak{a})$ is a prime ideal in $\mathbb{F}_q[x]/\langle x^n - \lambda \rangle$. Then $\Psi_1^i(\mathfrak{a}) = \langle f_1(x) \rangle$, where $f_1(x)$ is some monic irreducible divisor of $x^n - \lambda$. By Lemma 6.3, Hensel's Lemma, we have that there exists a polynomial $f_i(x)$ in $\mathcal{A}_i[x]$ that is a monic irreducible divisor of $x^n - \lambda$ in $\mathcal{A}_i[x]$ with $\Psi_1^i(f_i(x)) = f_1(x)$. Then by Lemma 6.4, we know that $\gamma \in \mathfrak{a}$. Then we have that $\langle f_i(x), \gamma \rangle \subseteq \mathfrak{a}$. The ring $(\mathcal{A}_i[x]/\langle x^n - \lambda \rangle)/\langle f_i(x), \gamma \rangle$ is a field, and therefore the ideal $\langle f_i(x), \gamma \rangle$ is maximal which gives that $\langle f_i(x), \gamma \rangle = \mathfrak{a}$.   □

For the infinite ring, we cannot guarantee that γ is in the ideal, but a similar proof gives the following.

Theorem 6.7 *The prime ideals in $\mathcal{A}_\infty[x]/\langle x^n - \lambda \rangle$ are either of the form $\langle f(x), \gamma^i \rangle$ or $\langle f(x) \rangle$, where $f(x)$ is any monic irreducible divisor of $x^n - \lambda$ over $\mathcal{A}_\infty$.*

Recall that a nonzero ideal $\mathfrak{a} \subseteq \mathcal{A}_i$ is called a primary ideal if $\mathfrak{a} \neq \mathcal{A}_i$ and whenever $ab \in \mathfrak{a}$, then either $a \in \mathfrak{a}$ or $b^k \in \mathfrak{a}$ for some positive integer k. A polynomial $f(x) \in \mathcal{A}_i[x]$ is primary if $\langle f(x) \rangle$ is a primary ideal of $\mathcal{A}_i[x]$.

Theorem 6.8 *Let i be a positive integer, the primary ideals in $\mathcal{A}_i[x]/\langle x^n - \lambda \rangle$ are $\langle f_i(x), \gamma^l \rangle$, where $\langle f_i(x) \rangle$ is an irreducible divisor of $x^n - \lambda$ over $\mathcal{A}_i$ and $0 \leq l < i$.*

Proof Let $\mathfrak{a} = \langle f_i(x), \gamma \rangle = \langle e_{\mathfrak{a}}(x), \gamma \rangle$ be an arbitrary prime ideal. It follows that $\mathfrak{a} = \langle e_{\mathfrak{a}}(x), \gamma \rangle = \langle e_{\mathfrak{a}}(x) \rangle + \langle \gamma \rangle$. Then, for any $0 \leq l < i$, and $a \in \mathfrak{a}^l$, there exist $a_{s_1},\cdots, a_{s_l} \in \mathfrak{a}$ such that

$$a = \sum_{s_1 \cdots s_l} a_{s_1} \cdots a_{s_l} = \sum_{s_1 \cdots s_l} \prod_{t=1}^{l} (e_{\mathfrak{a}}(x) y_t^{s_t} + \gamma z_t^{s_t})$$

$$= \sum_{s_1 \cdots s_l} (e_{\mathfrak{a}}(x) w_{s_1 \cdots s_l} + \gamma^l w'_{s_1 \cdots s_l})$$

$$= e_{\mathfrak{a}}(x) \sum_{s_1 \cdots s_l} w_{s_1 \cdots s_l} + \gamma^l \sum_{s_1 \cdots s_l} w'_{s_1 \cdots s_l} \in \langle e_{\mathfrak{a}}(x), \gamma^l \rangle.$$

This implies that $\mathfrak{a}^l \subseteq \langle e_{\mathfrak{a}}(x), \gamma^l \rangle$.

For the other direction, since $e_{\mathfrak{a}}(x) = e_{\mathfrak{a}}^l(x) \in \mathfrak{a}^l$ and $\gamma^l \in \mathfrak{a}^l$, we have that $\langle e_{\mathfrak{a}}(x), \gamma^l \rangle \subseteq \mathfrak{a}^l$. Hence $\mathfrak{a}^l = \langle e_{\mathfrak{a}}(x), \gamma^l \rangle$.

Let $\mathfrak{b}$ be an arbitrary primary ideal whose associated prime ideal is $\mathfrak{a} = \langle e_{\mathfrak{a}}(x), \gamma \rangle$, then (by [28], p. 200, Ex. 2) there exists an integer k such that $\mathfrak{a}^k \subseteq \mathfrak{b} \subseteq \mathfrak{a}$, and from this we get that $\mathfrak{a} = \mathfrak{a}^l$ for some l. Hence the results hold.   □

A similar proof gives the following.

Theorem 6.9 *The primary ideals in $\mathcal{A}_\infty[x]/\langle x^n - 1 \rangle$ are $\langle f_i(x) \rangle$ and $\langle f_i(x), \gamma^l \rangle$, where $\langle f_i(x) \rangle$ is an irreducible divisor of $x^n - 1$ over $\mathcal{A}_\infty$ and $0 \le l < \infty$.*

When $i < \infty$ we have that $\langle f_i(x), \gamma \rangle^i = \langle f_i(x) \rangle$. Then we have the following chain:

$$\langle f_i(x) \rangle \subseteq \langle f_i(x), \gamma^{i-1} \rangle \subseteq \cdots \subseteq \langle f_i(x), \gamma^2 \rangle \subseteq \langle f_i(x), \gamma \rangle. \tag{6.16}$$

In $\mathcal{A}_\infty[x]/\langle x^n - 1 \rangle$, we have the following infinite chain:

$$\langle f_\infty(x) \rangle \subseteq \langle f_\infty(x), \gamma^{i-1} \rangle \subseteq \cdots \subseteq \langle f_\infty(x), \gamma^2 \rangle \subseteq \langle f_\infty(x), \gamma \rangle. \tag{6.17}$$

Corollary 6.2 *Let i be any positive integer or infinity. Let $f_i^l(x), 1 \le l \le s, i \in \mathbb{N}$, denote the distinct monic irreducible divisors of $x^n - \lambda$ over $\mathcal{A}_i$. Then any ideal in $\mathcal{A}_i[x]/\langle x^n - \lambda \rangle$ can be written uniquely as*

$$\mathfrak{a} = \prod_{l=1}^{s} \langle f_i^l(x), \gamma \rangle^{m_l}, \tag{6.18}$$

where $0 \le m_l \le i$. If i is finite, then there are $(i+1)^s$ distinct ideals.

Proof The result follows from Theorem 6.8 and the Lasker-Noether decomposition theorem ([28], p. 209).   □

This takes us to our main theorem.

Theorem 6.10 *We have the following characterization of ideals in $\mathcal{A}_i[x]/\langle x^n - \lambda \rangle$ and $\mathcal{A}_\infty[x]/\langle x^n - \lambda \rangle$.*

- Let i be a positive integer, then any ideal of $\mathcal{A}_i[x]/\langle x^n - \lambda \rangle$ has the form

$$\langle f_0(x), \gamma f_1(x), \cdots, \gamma^{i-1} f_{i-1}(x) \rangle, \tag{6.19}$$

where $f_l(x)$ are divisors of $x^n - \lambda$ and $f_{i-1}(x) \mid \cdots \mid f_1(x) \mid f_0(x)$.
- Any ideal of $\mathcal{A}_\infty[x]/\langle x^n - \lambda \rangle$ has the form

$$\langle \gamma^{s_0} f_0(x), \gamma^{s_1} f_1(x), \cdots, \gamma^{s_{b-1}} f_{b-1}(x) \rangle, \tag{6.20}$$

where $0 \le s_0 < s_1 < \cdots < s_{b-1}$ for some b and $f_{b-1}(x) \mid \cdots \mid f_1(x) \mid f_0(x)$.

Proof The result follows from Theorem 6.8 and Corollary 6.2. $\qquad \square$

This result gives the form of constacyclic codes over any chain ring and simultaneously over any principal ideal ring. This follows from the fact that any principal ideal ring can be decomposed into the product of chain rings via the Chinese Remainder Theorem. The following corollary is a direct consequence of this.

Corollary 6.3 *Let R be a finite principal ideal ring, with R_i a finite chain ring with $R = CRT(R_1, R_2, \ldots, R_s)$. Then any ideal in $R[x]/\langle x^n - \lambda \rangle$ is of the form*

$$\mathfrak{a} = CRT(\mathfrak{a}_1, \mathfrak{a}_2, \ldots, \mathfrak{a}_s),$$

where $\mathfrak{a}_i$ is an ideal in $R_i[x]/\langle x^n - \lambda_i \rangle$ and $\lambda = CRT(\lambda_1, \lambda_2, \ldots, \lambda_s)$.

Proof Follows from Theorem 6.2. $\qquad \square$

Therefore, the characterization of ideals in Theorem 6.10 gives a characterization of all constacyclic codes over finite principal ideal rings.

In a similar manner, determining the structure of ideals in $R[x]/\langle x^n - \lambda \rangle$ where R is a finite commutative Frobenius local ring would find all constacyclic codes, which leads to our next open question.

Question 6.2 Classify all ideals $R[x]/\langle x^n - \lambda \rangle$ where R is a finite commutative Frobenius local ring.

6.3 Codes as Ideals in Group Rings

In the next three sections, we shall give some other generalizations of cyclic codes. In this section, we shall give an alternate generalization of the concept of cyclic codes and generalize this concept to groups other than the cyclic group.

It is natural to think of a cyclic code as an ideal in RC_n where C_n is the cyclic group of order n. As such, it seems that this should be generalized to consider codes as ideals in the group ring. The largest benefit in this is that you have a canonical subgroup of the automorphism group, namely the group G in the group ring RG.

See Theorem 6.12 for a complete description of this fact. We begin with the standard definition of a group ring.

Definition 6.2 Let G be a finite group of order n, then the group ring RG consists of elements of the form $\sum_{i=1}^{n} \alpha_i g_i$ where $\alpha_i \in R$ and $g_i \in G$. Addition in the group ring is done by coordinate addition, that is $\sum_{i=1}^{n} \alpha_i g_i + \sum_{i=1}^{n} \beta_i g_i = \sum_{i=1}^{n} (\alpha_i + \beta_i) g_i$. The product is given by

$$(\sum_{i=1}^{n} \alpha_i g_i)(\sum_{j=1}^{n} \beta_j g_j) = \sum_{i,j} \alpha_i \beta_j g_i g_j,$$

which gives that the coefficient of g_i in the product is $\sum_{g_j g_k = g_i} \alpha_i \beta_j$.

In this definition, if the ring is a field, then the object is a called a group algebra. Group rings are a well studied object and there is no restriction on the size of the group and ring. However, for our purposes we will always assume that G is a finite group and that R is a finite commutative Frobenius ring.

We shall now describe a construction of codes in a group algebra first appearing in [19]. It was expanded to group rings in [11].

Let R be a finite commutative Frobenius ring and let $G = \{g_1, g_2, \ldots, g_n\}$ be a group of order n. Let $v \in RG$ and define the matrix $\sigma(v)$ to be

$$\sigma(v) = \begin{pmatrix} \alpha_{g_1^{-1} g_1} & \alpha_{g_1^{-1} g_2} & \alpha_{g_1^{-1} g_3} & \cdots & \alpha_{g_1^{-1} g_n} \\ \alpha_{g_2^{-1} g_1} & \alpha_{g_2^{-1} g_2} & \alpha_{g_2^{-1} g_3} & \cdots & \alpha_{g_2^{-1} g_n} \\ \vdots & \vdots & \vdots & \vdots & \vdots \\ \alpha_{g_n^{-1} g_1} & \alpha_{g_n^{-1} g_2} & \alpha_{g_n^{-1} g_3} & \cdots & \alpha_{g_n^{-1} g_n} \end{pmatrix}. \tag{6.21}$$

From this matrix we are able to define a code of length n over R, specifically we define

$$C(v) = \langle \sigma(v) \rangle. \tag{6.22}$$

Given an element $v \in RG$, we have a matrix $\sigma(v) \in M_n(R)$ and a code $C(v)$ of length n over R. The following theorem first appeared in [11].

Theorem 6.11 *Let R be a finite commutative Frobenius ring and G a finite group of order n. For an element $v \in RG$, let $C(v)$ be the corresponding code in R^n, let $I(v)$ be the set of elements of RG such that $\sum \alpha_i g_i \in I(v)$ if and only if $(\alpha_1, \alpha_2, \ldots, \alpha_n) \in C(v)$. Then $I(v)$ is a left ideal in RG.*

Proof The rows in the matrix $\sigma(v)$ are vectors that correspond to the elements hv in RG where h is an arbitrary element of the group G. It is immediate that $I(v)$ is closed under addition since the sum of any two elements in $I(v)$ corresponds to the sum of the corresponding elements in $C(v)$.

We shall show that the product of an element in RG and an element in $I(v)$ is in $I(v)$. Let $w_1 = \sum \beta_i g_i$ be an arbitrary element in the group ring RG. Let

w_2 correspond to a vector in $C(v)$, which is of the form $\sum \gamma_j h_j v$. Then $w_1 w_2 = \sum \beta_i g_i \sum \gamma_i h_i v = \sum \beta_i \gamma_j g_i h_j v$, which corresponds to an element in $C(v)$ and gives that the element is in $I(v)$. Therefore, we have that $I(V)$ is a left ideal of RG. □

It is natural then to consider the ideals of a group ring as a class of interesting codes of length n over R.

Theorem 6.12 *Let R be a finite commutative Frobenius ring and G a finite group of order n. Let $v \in RG$ and $C(v)$ be the corresponding code in R^n. The automorphism group of $C(v)$ has a subgroup isomorphic to G.*

Proof We have by Theorem 6.11 that $I(v)$ is an ideal of RG. As such it is invariant by multiplication by elements of G which corresponds to the group acting on the coordinates of $C(v)$. Therefore, we have that the automorphism group of $C(v)$ contains G as a subgroup. □

This theorem shows the utility of this technique as one may start with a group and easily define a code over a finite commutative Frobenius ring with that group as part of the automorphism group. This is a natural generalization of the technique of generating a cyclic code by taking cyclic shifts of a vector to generate a cyclic code.

Question 6.3 For a given finite group G and a finite commutative Frobenius ring R, find the structure of all ideals in RG.

There are some natural choices of G which are quite close to the cyclic group. For example, the dihedral group would be an interesting choice. This group was used in [22] to construct the [48, 24, 12] extremal code. It would seem that dihedral codes could find as many engineering applications and in applications in other branches of mathematics as cyclic codes have found. In general, it would be of great interest to find finite groups that provide such applications and study codes in that particular group ring.

6.4 Quasicyclic Codes

Quasicyclic codes are another generalization of cyclic codes. They have received less attention than many generalizations, since they do not have a canonical algebraic description as ideals in a polynomial ring. In this section, we shall give a natural way to think about quasicyclic codes over finite fields by examining cyclic codes over the finite commutative Frobenius ring $R_{q,\Delta}$.

Let π be the standard cyclic shift. Specifically,

$$\pi(c_0, c_1, \ldots, c_{n-1}) = (c_{n-1}, c_0, c_1, \ldots, c_{n-2}).$$

We have the following definition.

Definition 6.3 Let C be a code over a finite commutative Frobenius ring R. Then if $\pi^k(C) = C$, we say that C is a quasicyclic code of index k.

Recall the definition of the ring $R_{q,\Delta}$ given in Eq. 4.21 and the corresponding Gray map Ψ. We shall generalize some of the results that first appeared for the binary case in [10].

From the definition of Ψ, we see that if $\mathbf{v}$ is a vector in R_Δ^n, with corresponding Gray map Ψ, then we have that $\Psi(\pi(\mathbf{v})) = \pi^\Delta(\Psi(\mathbf{v}))$. The next theorem, which gives a construction of quasicyclic codes of arbitrary index, follows immediately from this fact.

Theorem 6.13 *Let C be a linear cyclic code over the ring $R_{q,\Delta}$ of length n. Then $\Psi(C)$ is a linear quasicyclic code over $\mathbb{F}_q$ of length Δn and index Δ.*

Proof The code C is a cyclic code, so we have that $\pi(C) = C$. Then we have that $\Psi(C) = \Psi(\pi(C)) = \pi^\Delta(\Psi(C))$. Therefore, $\Psi(C)$ is a quasicyclic code of index Δ. $\square$

To study quasicyclic codes, it is then necessary to understand cyclic codes over $R_{q,\Delta}$. Therefore, we shall study the ideal structure of the corresponding polynomial ring.

Let A_Δ be the set of all monomials of $R_{q,\Delta}$ and let $\widehat{A}_\Delta$ be the subset of A_Δ of all monomials with one indeterminate. It is a simple computation to see that $|A_\Delta| = p_1^{k_1} p_2^{k_2} \cdots p_t^{k_t} = \Delta$ and $|\widehat{A}_\Delta| = p_1^{k_1} + p_2^{k_2} + \cdots + p_t^{k_t}$.

Any indeterminate $u_{p_i,j}$ may have an exponent in the set $J_i = \{0, 1, \ldots, p_i - 1\}$. Let $\alpha_i \in J_i^{k_i}$ and denote $u_{p_i,1}^{\alpha_i,1} \cdots u_{p_i,k_i}^{\alpha_i,k_i}$ by $u_i^{\alpha_i}$. Let $J = J_1^{k_1} \times \cdots \times J_t^{k_t}$. Each element $a \in A_\Delta$ can be written as $a = u^\alpha$ for some $\alpha \in J$, as the subset $\{u_{p_i,j}^{\alpha_{i,j}} | \alpha_{i,j} \neq 0\}_{(1 \leq i \leq t, 1 \leq j \leq k_i)} \subseteq \widehat{A}_\Delta$. Denote by $\widehat{a}$ the corresponding subset of $\widehat{A}_\Delta$.

Take the vector of exponents $\alpha = (\alpha_{1,1}, \ldots, \alpha_{1,k_1}, \ldots, \alpha_{t,1}, \ldots, \alpha_{t,k_t}) \in J$ and denote by $\bar{\alpha}$ the vector $(p_1 - \alpha_{1,1}, \cdots, p_1 - \alpha_{1,k_1}, \cdots, p_t - \alpha_{t,k_t})$. We note that $\bar{\bar{\alpha}} = \alpha$.

Denote by I_α the ideal $I_\alpha = \langle u^\alpha \rangle$, for $\alpha \in J$. Define $I_{(p_1, \cdots, p_1, p_2 \cdots, p_t, \cdots, p_t)} = \{0\}$. Define the ideal

$$\widehat{I}_\alpha = \langle u^{\widehat{\alpha}} \rangle = \langle u_{p_i,j}^{\alpha_{i,j}} | \alpha_{i,j} \neq 0 \rangle_{(1 \leq i \leq t, 1 \leq j \leq k_i)}. \tag{6.23}$$

The following theorems first appear in the binary case in [12].

Theorem 6.14 *Let $\alpha \in J$. Then the ideal $\widehat{I}_\alpha^\perp = I_{\bar{\alpha}}$.*

Proof It is easy to see that $I_{\bar{\alpha}} \subset \widehat{I}_\alpha^\perp$.

Suppose it is not true that $\widehat{I}_\alpha^\perp \subset I_{\bar{\alpha}}$. Then there exist an element $b = \sum_{\beta \in J} c_\beta u^\beta \in \widehat{I}_\alpha^\perp$ that does not belong to $I_{\bar{\alpha}}$. It follows that there exists a particular β such that $c_\beta \neq 0$ and $\beta_{i,j} < \bar{\alpha}_{i,j}$ for some i and j. Then, $u_{p_i,j}^{\alpha_{i,j}} \cdot b \neq 0$ for $u_{p_i,j}^{\alpha_{i,j}} \in \widehat{I}_\alpha$, which gives $b \notin \widehat{I}_\alpha^\perp$ and $\widehat{I}_\alpha^\perp \subset I_{\bar{\alpha}}$. $\square$

Here, we have $\widehat{I_{\bar{0}}^{\perp}} = R_{\Delta}^{\perp} = \{0\} = I_{(p_1,\cdots,p_1,p_2\cdots,p_t,\cdots,p_t)} = I_{\bar{0}}$.

Theorem 6.15 *The number of elements of I_{α} is $q^{\prod_{i\in\bar{\alpha}} i}$ and the number of elements of $\widehat{I_{\alpha}}$ is $q^{\Delta-\prod_{i\in\alpha} i}$.*

Proof There are $p_1 - \alpha_{1,1}$ different monomials fixing all the indeterminates except the first one, $u_{p_1,1}$ in the set of all monomials of I_{α}. There are $p_1 - \alpha_{1,2}$ different monomials fixing all the indeterminates except the second one, $u_{p_1,2}$. By induction, there are $\prod_{1\leq i\leq t, 1\leq j\leq k_i}(p_i - \alpha_{i,j})$ different monomials in I_{α}.

Then, since $\bar{\alpha}$ is the vector $(p_1 - \alpha_{1,1}, \cdots, p_1 - \alpha_{1,k_1}, \cdots, p_t - \alpha_{t,k_t})$ and all elements in I_{α} are a linear combination of its monomials, we have that $|I_{\alpha}| = q^{\prod_{i\in\bar{\alpha}} i}$. By Theorem 6.14, it follows that $|\widehat{I_{\alpha}}| = q^{\Delta-\prod_{i\in\alpha} i}$.

Since cyclic codes over $R_{q,\Delta}$ can be viewed as ideals in $R_{q,\Delta}[x]/\langle x^n - 1\rangle$, we use the canonical decomposition of rings to obtain the following theorem, noting that when the characteristic of $\mathbb{F}_q$ and n are relatively prime the factorization is unique.

Theorem 6.16 *Let n be an integer relatively prime to the characteristic of $\mathbb{F}_q$ and $x^n - 1 = f_1 f_2 \ldots f_r$. The ideals in $R_{q,\Delta}[x]/\langle x^n - 1\rangle$ can be written as $I \cong I_1 \oplus I_2 \oplus \cdots \oplus I_r$ where I_i is an ideal of the ring $R_{q,\Delta}[x]/\langle f_i\rangle$, for $i = 1, \ldots, r$.*

We shall now examine what form these ideals take. We generalize the approach used for the binary case in [12]. Let f be an irreducible polynomial in $\mathbb{F}_q[x]$, then f is a basic monic irreducible polynomial over $R_{q,\Delta}$.

Since the polynomial is irreducible, we have that $\mathbb{F}_q[x]/\langle f\rangle$ is a finite field of order $q^{\deg(f)}$. Let $L_{0,0} = \mathbb{F}_q[x]/\langle f\rangle$ and $L_{p_1,1} = L_{0,0}[u_{p_1,1}]/\langle u_{p_1,1}^{p_1}\rangle$. For $1 \leq i \leq t, 1 \leq j \leq k_i$, define

$$L_{p_i,j} = \begin{cases} L_{p_{i-1},k_{i-1}}[u_{p_i,1}]/\langle u_{p_i,1}^{p_i}\rangle & \text{if } j = 1, \\ L_{p_i,j-1}[u_{p_i,j}]/\langle u_{p_i,j}^{p_i}\rangle & \text{otherwise.} \end{cases} \qquad (6.24)$$

Any element $a \in L_{p_i,j}$ can be written as

$$a = a_0 + a_1 u_{p_i,j} + a_2 u_{p_i,j}^2 + \cdots + a_{p_i-1} u_{p_i,j}^{p_i-1}$$

where $a_0, \ldots, a_{p_i-1}$ belong to $L_{p_i,j-1}$ if $j \neq 1$ or to $L_{p_{i-1},k_{i-1}}$ if $j = 1$.

Theorem 6.17 *Let $a = \sum_{d=0}^{p_i-1} a_d u_{p_i,j}^d$ be an element of $L_{p_i,j}$. The element a is a unit in $L_{p_i,j}$ if and only if a_0 is a unit in $L_{p_i,j-1}$ if $j \neq 1$ or in $L_{p_{i-1},k_{i-1}}$ if $j = 1$.*

Proof Assume that a_0 is a unit in $L_{p_i,j-1}$ if $j \neq 1$ or in $L_{p_{i-1},k_{i-1}}$ if $j = 1$. Define $b = a_0^{-1}(\sum_{d=1}^{p_i-1} a_d u_{p_i,j}^d)$. Then b is a zero divisor and $1 + b$ is a unit since $(1 + b)(1 + b + b^2 + \cdots + b^{p_i-1}) = 1$. This gives that $a_0(1 + b) = a$ is also a unit.

If the element a_0 is not a unit, then there exists b in $L_{p_i,j-1}$ if $j \neq 1$ or in $L_{p_{i-1},k_{i-1}}$ if $j = 1$, such that $ba_0 = 0$. Therefore, $bu_{p_i,j}^{p_i-1}a = 0$. $\qquad\square$

Denote the group of units of $L_{p_i,j}$ by $\mathcal{U}(L_{p_i,j})$. By the previous result we can see that

$$|\mathcal{U}(L_{p_i,j})| = \begin{cases} |\mathcal{U}(L_{p_{i-1},k_{i-1}})||L_{p_{i-1},k_{i-1}}| & \text{if } j = 1, \\ |\mathcal{U}(L_{p_i,j-1})||L_{p_i,j-1}| & \text{otherwise.} \end{cases} \qquad (6.25)$$

We have that $|\mathcal{U}(L_{p_1,1})| = q^{\deg(f)}(q^{\deg(f)} - 1)$ since we have that $|\mathcal{U}(L_{0,0})| = q^{\deg(f)} - 1$. Then by induction, we have that

$$|L_{p_t,k_t}| = (q^{\deg(f)})^\Delta \text{ and } |\mathcal{U}(L_{p_t,k_t})| = (q^{\deg(f)})^\Delta - (q^{\deg(f)})^{\Delta-1}. \qquad (6.26)$$

Theorem 6.18 *The ideals of L_{p_t,k_t} are in bijective correspondence with the ideals of $R_{q,\Delta}$.*

Proof Theorem 6.17 gives that the zero-divisors of L_{p_t,k_t} are of the form $\sum c_\alpha u_1^{\alpha_1} \cdots u_t^{\alpha_t}$ with $c_\alpha \in L_{0,0}$ and $c_0 = 0$ and that there are $(q^{\deg(f)})^{\Delta-1}$ of them. The result follows. $\square$

The following corollary is an immediate consequence of the previous theorem.

Corollary 6.4 *Let n be an integer relatively prime to the characteristic of $\mathbb{F}_q$. Let $x^n - 1 = f_1 f_2 \dots f_r$ be the unique factorization of $x^n - 1$ into basic irreducible polynomials over $R_{q,\Delta}$. Let I_Δ be the number of ideals in $R_{q,\Delta}$. Then the number of linear cyclic codes of length n over $R_{q,\Delta}$ is $(I_\Delta)^r$.*

Of course, it is not true that all quasicyclic codes over $\mathbb{F}_q$ are images, under the Gray map, of cyclic codes over $R_{q,\Delta}$. This fact leads to the following question which is couched in the broadest setting.

Question 6.4 Let R be a finite commutative Frobenius ring. Classify all quasicyclic codes with index k and length n over R.

6.5 θ-Cyclic Codes

In this section, we shall describe another generalization of cyclic codes. This is generalization has been extremely important in finding good codes, namely those codes with large minimum distances with respect to their ambient space and size. The theory of θ-cyclic codes over fields has been described in [1–5]. They have also been studied over rings in [18, 26] for example. We note that these codes are also often called skew cyclic codes.

Definition 6.4 Let R be a Frobenius ring and let θ be an automorphism of R. A θ-cyclic code is a linear code C such that

$$(c_0, c_1, \dots, c_{n-1}) \in C \Rightarrow (\theta(c_{n-1}), \theta(c_0), \theta(c_1), \dots, \theta(c_{n-2})) \in C.$$

We define the following skew polynomial ring. Let

$$R[x, \theta] = \{a_0 + a_1 x^1 + \cdots + a_{n-1} x^{n-1} \mid a_i \in R, n \in \mathbb{N}\},$$

where addition is the usual polynomial addition and multiplication is defined by $xa = \theta(a)x$ for all $a \in R$. We note that this ring is a non-commutative ring even though the code alphabet is commutative. This ring is a common example in non-commutative ring theory. We need to show what algebraic structures correspond to θ-cyclic codes. However, unlike our previous work, we must be sure to indicate whether it is a left or a right module.

Theorem 6.19 *Let C be a linear θ-cyclic code over a finite commutative Frobenius ring R. Let $\mathfrak{a}$ be the set of all polynomials of the form $a_0 + a_1 x + a_2 x^2 + \cdots + a_{n-1} x^{n-1}$ where $(a_0, a_1, \ldots, a_{n-1}) \in C$. Then $\mathfrak{a}$ is a left module of $R[x, \theta]/\langle x^n - 1 \rangle$.*

Proof It is clear that $\mathfrak{a}$ is an additive subgroup. Let $a(x) = a_0 + a_1 x + \ldots a_{n-1} x^{n-1} \in \mathfrak{a}$. Then we have that

$$\begin{aligned}
xa(x) &= xa_0 + xa_1 x + xa_2 x^2 + \cdots + xa_{n-1} x^{n-1} \\
&= \theta(a_0)x + \theta(a_1)x^2 + \cdots + \theta(a_{n-1})x^n \\
&= \theta(a_{n-1}) + \theta(a_0)x + \cdots + \theta(a_{n-2})x^{n-1} \in \mathfrak{a}.
\end{aligned}$$

Therefore, $\mathfrak{a}$ is a left module of $R[x, \theta]/\langle x^n - 1 \rangle$. $\square$

When the alphabet is a finite field, then finding left ideals is equivalent to finding right divisors of $x^n - 1$ in $R[x, \theta]$, see [6] for a complete description. See also [5], for a description of θ-cyclic codes where the alphabet is a Galois ring. In general, we have the following open question.

Question 6.5 Determine the structure of all θ-cyclic codes in $R[x, \theta]/\langle x^n - \lambda \rangle$ where R is a finite commutative Frobenius ring.

As an example of a ring that can be used in this setting, consider the ring $\mathbb{F}_2 + v\mathbb{F}_2$, where $v^2 = v$. This ring has an automorphism which interchanges v and $1 + v$. See [26] for a description of skew cyclic codes in this case.

References

1. Boucher, D., Ulmer, F.: Self-dual skew codes and factorization of skew polynomials. J. Symbolic Comput. **60**, 47–61 (2014)
2. Boucher, D., Ulmer, F.: Linear codes using skew polynomials with automorphisms and derivations. Des. Codes Crypt. **70**(3), 405–431 (2014)
3. Boucher, D., Ulmer, F.: Codes as modules over skew polynomial rings. In: Cryptography and Coding. Lecture Notes in Computer Science, vol. 5921, pp. 38–55 (2009)

4. Boucher, D., Ulmer, F.: Coding with skew polynomial rings. J. Sym. Comput. **44**(12), 1644–1656 (2009)
5. Boucher, D., Solé, P., Ulmer, F.: Skew constacyclic codes over Galois rings. Adv. Math. Comput. **2**(3), 273–292 (2008)
6. Boucher, D., Geiselmann, W., Ulmer, F.: Skew cyclic codes. AAECC **18**(4), 379–389 (2007)
7. Dinh, H., López-Permouth, S.R.: Cyclic and negacyclic codes over finite chain rings. IEEE Trans. Inf. Theory **50**, 1728–1744 (2004)
8. Calderbank, A.R., Sloane, N.J.A.: Modular and p-adic cyclic codes. Des. Codes Crypt. **6**, 21–35 (1995)
9. Dougherty, S.T., Fernández-Córdoba, C.: Kernels and ranks of cyclic and negacyclic quaternary codes. Des. Codes Crypt. **81**(2), 347–364 (2016). doi:10.1007/s10623-015-0163-6
10. Dougherty, S.T., Fernández-Córdoba, C., Ten-Valls, R.: Quasi-cyclic codes as cyclic codes over a family of local rings. Finite Fields Appl. **40**, 138–149 (2016)
11. Dougherty, S.T., Gildea, J., Taylor, R., Tylshchak, A.: Constructions of self-dual and formally self-dual codes from group rings (in submision)
12. Dougherty, S.T., Kaya, A., Saltürk, E.: Cyclic codes over local frobenius rings of order 16. Adv. Math. Comm. **11**(1), 99–114 (2017)
13. Dougherty, S.T., Ling, S.: Cyclic codes over $\mathbb{Z}_4$ of even length. Des. Codes Crypt. **39**(2), 127–153 (2006)
14. Dougherty, S.T., Liu, H.: Cyclic codes over formal power series. Acta Math. Sci. **31**(1), 331–343 (2010)
15. Dougherty, S. T., Liu, H.: Cyclic codes over formal power series rings. Acta Math. Sci. Ser. B Eng. Ed. **311**, 331–343 (2011)
16. Dougherty, S.T., Liu, H., Park, Y.H.: Lifted codes over finite chain rings. Math. J. Okayama Univ. **53**, 39–53 (2010)
17. Dougherty, S.T., Park, Y.H.: On modular cyclic codes. Finite Fields Appl. **13**(1), 31–57 (2007)
18. Ezerman, M., Ling, S., Solé, P., Yemen, O.: From skew-cyclic codes to asymmetric quantum codes. Adv. Math. Commun. **5**(1), 41–57 (2011)
19. Hurley, T.: Group rings and rings of matrices. Int. J. Pure Appl. Math. **31**(3), 319–335 (2006)
20. Kanwar, P., López-Permouth, S.R.: Cyclic codes over the integers modulo p^m. Finite Fields Appl. **3**, 334–352 (1997)
21. McDonald, B.R.: Finite Rings with Identity. Marcel Dekker Inc., New York (1974)
22. McLoughlin, I.: A group ring construction of the [48, 24, 12] Type II linear block code. Des. Codes Crypt. **63**, 29–41 (2012)
23. Norton, G.H., Sălăgean, A.: On the structure of linear and cyclic codes over a finite chain ring. Appl. Algebra Eng. Commun. Comput. **10**, 489–506 (2000)
24. Pless, V.S., Qian, Z.: Cyclic codes and quadratic residue codes over $\mathbb{Z}$. IEEE-IT **42**(5), 1594–1600 (1996)
25. Prange, E.: Cyclic error-correcting codes in two symbols. Technical Note TN-57-103, Air Force Cambridge Research Labs, Bedford Mass
26. Solé, P., Yemen, O.: Binary quasi-cyclic codes of index 2 and skew polynomial rings. Finite Fields Appl. **18**(4), 685–699 (2012)
27. Wan, Z.: Cyclic codes over Galois rings. Alg. Colloq. **6**, 291–304 (1999)
28. Zariski, O., Samuel, P.: Commutative Algebra. Van Nostrand, New York (1958)

Index

Symbols
A_k, 54
M-dual code, 78
R_k, 50
$R_{q,\Delta}$, 52, 98
S_k, 50
T_k, 51
Σ_{2k}, 65
Θ_{2k}, 65
θ-cyclic codes, 98
i-dual code, 78

A
Additive code, 33
Artinian ring, 15
Assmus, E.F., 2
Assmus-Mattson Theorem, 72

B
Bachoc weight, 43
Basis, 25
BIBD, 77
Blake, I. F., 2

C
Calderbank, A.R., 86
Chain ring, 14, 21, 39, 63
Character, 31
Character table, 17, 31
Chinese Remainder Theorem, 13, 18–22, 37, 38, 42, 60, 63, 84, 86
Code, 3
Complete weight enumerator, 30
Complex lattice, 66

Constacyclic, 83, 84
Construction A, 66
Coprime, 18
Cyclic, 83
Cyclic codes, 83

D
Delsarte, P. , 2
Design, 73
Direct product, 61
Dougherty, S.T., 86
Duality, 31
Dual lattice, 64

E
Epimorphism, 55
Equivalence, 7
Euclidean weight, 43

F
Formal power series, 86
Free code, 19, 61
Frobenius, 16
Frobenius ring, 15
Fundamental volume, 64

G
Galois ring, 14, 38
Generating character, 17, 37, 39
Generator matrix, 22
Gilbert-Varshamov construction, 26
Golay code, 9, 64, 69, 72, 84
Gray map, 41, 49, 50, 53, 96

© The Author(s) 2017
S.T. Dougherty, *Algebraic Coding Theory Over Finite Commutative Rings*,
SpringerBriefs in Mathematics, DOI 10.1007/978-3-319-59806-2

Printed in the United States
By Bookmasters